CHEMICAL ANALYSIS

A SERIES OF MONOGRAPHS ON
ANALYTICAL CHEMISTRY AND ITS APPLICATIONS

Editors

P. J. ELVING · I. M. KOLTHOFF

Advisory Board

Leslie S. Ettre
L. T. Hallett
Herman A. Liebhafsky
J. J. Lingane
W. Wayne Meinke
Louis Meites

J. Mitchell, Jr.
George H. Morrison
Charles N. Reilley
Anders Ringbom
E. B. Sandell

Donald T. Sawyer
C. A. Streuli
Fred Stross
E. H. Swift
Nobuyuki Tanaka
Wesley Wm. Wendlandt

VOLUME 17

WILEY-INTERSCIENCE

A Division of John Wiley & Sons, Inc.
New York/London/Sydney/Toronto

CHEMICAL ANALYSIS

Vol. 1 **The Analytical Chemistry of Industrial Poisons, Hazards, and Solvents.** Second Edition. By Morris B. Jacobs
Vol. 2. **Chromatographic Adsorption Analysis.** By Harold H. Strain (*out of print*)
Vol. 3. **Colorimetric Determination of Traces of Metals.** Third Edition. By E. B. Sandell
Vol. 4. **Organic Reagents Used in Gravimetric and Volumetric Analysis.** By John F. Flagg (*out of print*)
Vol. 5. **Aquametry: Application of the Karl Fischer Reagent to Quantitative Analyses Involving Water.** By John Mitchell, Jr. and Donald Milton Smith (*temporarily out of print*)
Vol. 6. **Analysis of Insecticides and Acaricides.** By Francis A. Gunther and Roger C. Blinn (*Out of print*)
Vol. 7. **Chemical Analysis of Industrial Solvents.** By Morris B. Jacobs and Leopold Scheflan
Vol. 8. **Colorimetric Determination of Nonmetals.** Edited by David F. Boltz
Vol. 9. **Analytical Chemistry of Titanium Metals and Compounds.** By Maurice Codell
Vol. 10. **The Chemical Analysis of Air Pollutants.** By Morris B. Jacobs
Vol. 11. **X-Ray Spectrochemical Analysis.** Second Edition. By L. S. Birks
Vol. 12. **Systematic Analysis of Surface-Active Agents.** Second Edition. By Milon J. Rosen and Henry A. Goldsmith
Vol. 13. **Alternating Current Polarography and Tensammetry.** By B. Breyer and H. H. Bauer
Vol. 14. **Flame Photometry.** By R. Herrmann and J. Alkemade
Vol. 15. **The Titration of Organic Compounds** (*in two parts*). By M. R. F. Ashworth
Vol. 16. **Complexation in Analytical Chemistry: A Guide for the Critical Selection of Analytical Methods Based on Complexation Reactions.** By Anders Ringbom
Vol. 17. **Electron Probe Microanalysis.** Second Edition. By L. S. Birks
Vol. 18. **Organic Complexing Reagents: Structure, Behavior, and Application to Inorganic Analysis.** By D. D. Perrin
Vol. 19. **Thermal Methods of Analysis.** By Wesley Wm. Wendlandt
Vol. 20. **Amperometric Titrations.** By John T. Stock
Vol. 21. **Reflectance Spectroscopy.** By Wesley Wm. Wendlandt and Harry G. Hecht
Vol. 22. **The Analytical Toxicology of Industrial Inorganic Poisons.** By the late Morris B. Jacobs
Vol. 23. **The Formation and Properties of Precipitates.** By Alan G. Walton
Vol. 24. **Kinetics in Analytical Chemistry.** By Harry B. Mark, Jr. and Garry A. Rechnitz
Vol. 25. **Atomic Absorption Spectroscopy.** By Walter Slavin
Vol. 26. **Characterization of Organometallic Compounds** (*in two parts*). Edited by Minoru Tsutsui
Vol. 27. **Rock and Mineral Analysis.** By John A. Maxwell
Vol. 28. **The Analytical Chemistry of Nitrogen and Its Compounds** (*in two parts*). Edited by C. A. Streuli and Philip R. Averell
Vol. 29. **The Analytical Chemistry of Sulfur and Its Compounds** (*in three parts*). By J. H. Karchmer
Vol. 30. **Ultramicro Elemental Analysis.** By Günther Tölg
Vol. 31. **Photometric Organic Analysis** (*in two parts*). By Eugene Sawicki
Vol. 32. **Determination of Organic Compounds: Methods and Procedures.** By Frederick T. Weiss
Vol. 33. **Masking and Demasking of Chemical Reactions.** By D. D. Perrin
Vol. 34. **Neutron Activation Analysis.** By D. De Soete, R. Gijbels, and J. Hoste
Vol. 35. **Laser Raman Spectroscopy.** By Marvin C. Tobin
Vol. 36. **Emission Spectrochemical Analysis.** By Morris Slavin

ELECTRON PROBE MICROANALYSIS

Second Edition

L. S. Birks

X-Ray Optics Branch
U.S. Naval Research Laboratory
Washington, D.C.

WILEY-INTERSCIENCE

A Division of John Wiley & Sons, Inc.
New York · London · Sydney · Toronto

Copyright © 1963, 1971, by John Wiley & Sons, Inc.

All rights reserved. Published simultaneously in Canada.

No part of this book may be reproduced by any means, nor transmitted, nor translated into a machine language without the written permission of the publisher.

Library of Congress Catalog Card Number: 74-165827

ISBN 0-471-07533-7

Printed in the United States of America.

10 9 8 7 6 5 4 3 2 1

PREFACE

In the eight years since the first edition of this book appeared electron probe analysis has matured into a well recognized and commonly accepted tool in nearly all fields of science. Commercial instruments have gone into second- and even third-generation designs with options such as scanning microscopy, multiple spectrometers, energy dispersive detectors, and computer coupling for automatic control and data treatment.

Spectacular advances have occurred both in qualitative analysis, in which scanning displays are now made with 100-Å diameter beams, and in quantitative analysis, in which correction procedures have become routine and inexpensive and are usually done by computer. Scanning displays are frequently prepared in full color by superposing displays for several elements through selected color filters (unfortunately, book printing does not lend itself readily to the reproduction of color prints as illustrations).

Because of the advances in the electron probe technique, the second edition represents a complete rewriting of much of the book. Although the electron probe has by no means achieved its ultimate capability, it does seem to have reached a temporary plateau. Hopefully the material presented here will give the analyst of the 1970s a starting basis for working in this important field.

L. S. Birks

Washington, D.C.
May 1971

PREFACE TO THE FIRST EDITION

One of the remarkable things about electron probe microanalysis is that it captured the imagination of so many analysts even during its initial period, when there were less than a dozen instruments in existence and very few actual applications published. The concept of analyzing micron-sized areas on specimens immediately indicated its usefulness, especially in metallurgy, where segregations and precipitates in that size range had long been observed by microscopy but had never been analyzed except on an average basis by x-ray diffraction or chemical analysis of residues. These, as well as a number of problems in corrosion, diffusion, phases in minerals, and localization of elements in biologicals, had been studied primarily by indirect methods, and merely awaited the proper tool to push the frontiers forward.

The electron probe concept was a natural outgrowth of (1) electron microscopy, which had introduced valuable techniques for forming and controlling electron beams, and (2) x-ray spectrochemical analysis, which had proved to be a simple and straightforward means of quantitative chemical analysis. By the late 1940's both of those fields had reached a "state of the art" that allowed dependable instruments to be constructed and allowed analysts to use them as routine tools. In 1947 James Hillier obtained a patent on a proposed device to combine a fine-focused beam of electrons to generate characteristic x-rays in a solid specimen and an x-ray spectrometer to measure the wavelength and intensity of the x-rays emitted. About 1949 Castaing and Guinier in France and, independently, Borovskii in Russia actually constructed such instruments. By 1955 a number of independent laboratories throughout Europe and the United States were designing and constructing probes for their own use, and manufacturers were considering commercial production of the instrument. Between 1955 and 1960 considerable interest was generated by lectures at scientific meetings and descriptive articles in the literature, but the number of instruments in operation remained about ten. During 1960 and 1961 increased commercial production and delivery of instruments was achieved, and the electron probe began to take its place as an important analytical tool.

It is the purpose in this book to look ahead to the immediate future as the number of analysts increases rapidly and brings about the need for an understanding of the concepts and principles by persons whose primary interest is not in x-ray or electron optics as such, but rather in using the electron probe as a tool in their research. In order to make the book short and readable, the attempt has been to explain and elucidate the subject so that metallurgists, chemists, mineralogists, biologists, etc., will be able to take advantage of the technique without becoming involved in instrument design or unnecessary theory. It has appeared desirable to include sections on certain instrumental components and principles, however, because they are necessary to a proper understanding of the limitations as well as the possibilities of the instrument. Because of the importance of quantitative analysis and the difficulty of cross-checking the results by more standard techniques, considerable space has been devoted to explaining how quantitative analysis is accomplished.

There is likely to be some disagreement on emphasis, depending on the background and interest of the particular reader, and certainly this book should not be considered as a complete reference volume for all questions concerning electron probe analysis. It is hoped that it will be a useful guide to those entering the field who may not be aware of all the ramifications and interrelations in chemical analysis by x-rays. Pertinent literature references are cited and the analyst should build up a library of specific papers relating to his individual interests.

I wish to express my thanks to my colleagues at the U.S. Naval Research Laboratory, particularly to R. E. Seebold, for their aid in proofreading the manuscript and making helpful suggestions.

L. S. BIRKS

U.S. Naval Research Laboratory
Washington, D.C.

CONTENTS

CHAPTER

1	**Introduction**	1
	1.1 Background	1
	1.2 The Instrument	1
	1.3 Limitations on Atomic Numbers and Detectability	3
	1.4 Purpose of this Book	4
2	**Historical Background**	5
	2.1 X-Ray Spectroscopy	5
	2.2 Electron Microscopy	6
	2.3 The Electron Probe Concept	7
	2.4 Development of a Variety of Instruments	9
	2.5 Electron Probe Philosophies	18
	2.6 Future Operators of Electron Probes	18
3	**Electron Optics Column and Circuits**	20
	3.1 The Vacuum System	20
	3.2 Electron Gun	21
	3.3 Voltage and Current	23
	3.4 Electron Lens Systems	23
	3.5 Combination Electron Probe–Electron Microscope	29
	3.6 Beam Deflection and Scanning	30
	3.7 Electron Lens Aberrations	31
	3.8 Alignment of Electron Optics Column	37
4	**X-Ray Spectrometers**	41
	4.1 Bragg Diffraction	41
	4.2 Crystal Parameters	42
	4.3 Curved-Crystal Optics	45
	4.4 Fanning Divergence	47
	4.5 Intensity of Diffraction Lines	50
	4.6 Resolution	52
	4.7 Line/Background Ratio	53

ix

	4.8	Mechanics of Spectrometers	54
	4.9	Limitations on Specimen Position and Scanning Image Size	55
5		**Detectors and Energy Dispersion**	57
	5.1	X-Ray Detectors	57
	5.2	Electron Detectors	64
	5.3	Luminescence Detectors	66
	5.4	Energy Dispersion	66
6		**Types of Specimens, Preparation, Examination, and Interpretation**	71
	6.1	Categories of Specimens	71
	6.2	Factors in Specimen Preparation	71
	6.3	Hard Materials	71
	6.4	Soft Materials	75
	6.5	Inclusions and Precipitates	80
	6.6	Pigments and Particulate Matter	86
	6.7	Diffusion Zones	87
	6.8	Thin-Film Specimens	92
	6.9	Solid State Devices	95
	6.10	Elements of Atomic Number 11 and Below	98
	6.11	Miscellaneous Techniques in Displaying and Interpreting Specimen Information	99
7		**Introduction to Quantitative Analysis: Empirical Methods; The Correction-Factor Approach**	101
		Empirical Methods	
	7.1	Calibration Curves	101
	7.2	Comparison Standards	102
	7.3	Empirical Coefficient Equations	104
	7.4	Using the Coefficients in Analysis	104
		The Correction-Factor Approach	
	7.5	Electron-Beam Penetration and Backscatter	106
	7.6	Electron Generation of Characteristic X-Rays	109
	7.7	Secondary Fluorescence	112
	7.8	The Correction Factor Equation	112
	7.9	The Absorption Correction Factor, A	113
	7.10	The Atomic Number Correction Factor, Z	113
	7.11	The Characteristic Fluorescence Factor, F	116

	7.12 Numerical Example of the Correction Factor Equation	116
	7.13 Computer Programs and Iteration	119
	7.14 Errors Due to Particle Size or Sharp Boundaries	120
8	**Advanced Computer Methods for Quantitative Analysis**	**124**
	8.1 The Electron-Transport Concept	125
	8.2 Ramifications and Limitations of the Transport Approach	128
	8.3 Monte Carlo Methods	129
9	**Related Instrumentation and Techniques**	**130**
	9.1 The Ion Probe Microanalyzer	130
	9.2 Electron Spectrometry	133
	9.3 Scanning Electron Microscopy	135
	9.4 Electron and X-Ray Diffraction	136
	9.5 Divergent-Beam Patterns	140
	9.6 X-Ray Microscopy	144
	9.7 Combined Electron Probe-Electron Microscope	145

APPENDICES

1	Tables of Mass Attenuation Coefficients for $K\alpha$ and $L\alpha_1$ Lines	147
2	Average Values of Fluorescent Yields	172
3	Jump Factors for Selected Elements	174
4	Excitation Energy for K-, L-, and M-Series	176
5	Characteristic Wavelengths of the K, L, and M Series Lines	179
6	Useful Analyzer Crystals and Their Spacings	182
References		183
Author Index		187
Subject Index		189

Electron Probe Microanalysis

CHAPTER

1

INTRODUCTION

1.1 BACKGROUND

Electron probe analysis provides a means for studying the local composition and structure of hetrogeneous materials such as alloys, thin films, minerals, ceramics, biological specimens, solid-state circuits, and so on. It does this by irradiating the sample with an electron beam focused to about 1 μm diameter at the surface of the sample and by measuring the characteristic x-rays or the electrons emerging from the irradiated area. The characteristic x-rays allow qualitative or quantitative analysis of chemical composition of precipitates, diffusion zones, or local enrichment. Backscattered electrons distinguish variations in physical topography (by scanning electron microscopy) variations in electric field (solid-state circuits) or variations in average atomic number. Transmitted electrons may be used in combination electron probe–electron microscope instruments.

Electron probe analysis is a powerful tool because the properties and reactions of most materials depend not only on the average composition but particularly on the localization of elements within the material. It is fortutious that the degree of localization is often on the micron-size scale which is about the practical limit for the effective x-ray source size even though the electron beam may be focused to a smaller diameter.

1.2 THE INSTRUMENT

Figure 1.1 shows the components of the electron probe schematically. They are as follows.

1. An electron optical system consisting of an electron gun operated at selected potentials from about 5 to 50 keV and stabilized to 1 part in 10^4 to maintain the proper focusing of the beam at the specimen; electron lenses to demagnify the image of the filament so as to achieve a beam diameter of 0.1 to 3 or 4 μm at the sample; and deflection plates or coils to sweep the focused beam back and forth over and area of the sample. This, when coupled with the detector readout, allows an

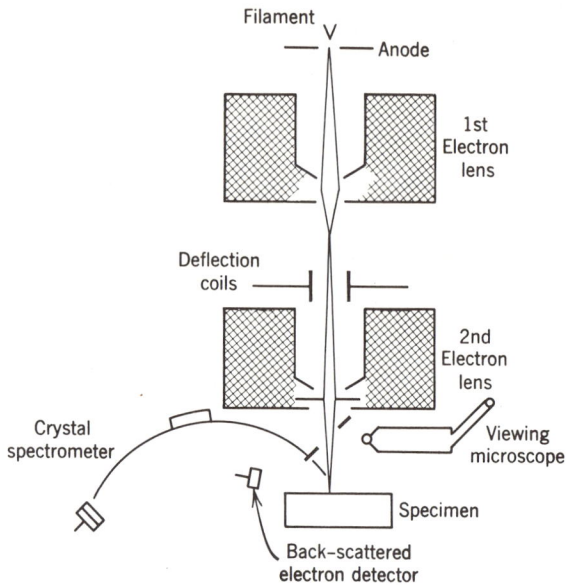

Fig. 1.1 Schematic diagram of electron probe microanalyzer.

x-ray or electron picture of the sample to be displayed on a cathode ray tube as shown in Figure 6.2.

2. A visual viewing system, usually a microscope, to allow the operator to select the general area of the sample to be examined. Electroluminescence may be observed or photographed through the microscope or a phototube output may be used for a display picture similar to the x-ray or electron displays.

3. Electron detectors for emitted electrons or specimen current to ground. The emitted electron detectors may be used without accelerating or retarding grids to measure low-energy (secondary electron emission) or high-energy (backscattered) electrons.

4. Total x-ray detectors such as solid-state detectors for energy dispersion, or x-ray spectrometers for high-resolution selection of characteristic x-rays.

The electron optical system, specimen, and electron detectors must be in a high-vacuum system pumped by a diffusion pump. X-ray optics may be in air or vacuum depending on the wavelengths to be measured. The viewing system may be in air or vacuum but must, of course, emerge to air for the operator unless a television type of pickup is used. Instrumentation will be discussed in more detail in Chapters 3, 4, and 5.

1.3 LIMITATIONS ON ATOMIC NUMBERS AND DETECTABILITY

Electron probe analysis has approximately the same limitations as fluorescent x-ray spectrometry as far as the elements which can be measured are concerned. That is, elements of atomic number greater than 22 (titanium) can be measured with air-path x-ray optics; elements from 11 (sodium) to 22 (titanium) can be measured with vacuum or helium-path x-ray optics; elements below 11 (sodium) and down to 5 (boron) require high vacuum and special crystals and detectors; elements below boron have been measured under laboratory conditions but ordinarily are not feasible with commercial electron probes.

The limit of detectability is not as low with electron probes as with fluorescent x-ray spectroscopy because of the relatively higher background intensity. This higher background intensity with electron excitation arises from the fact that electrons generate a strong, continuous x-ray spectrum as well as the characteristic lines from the elements. With primary x-ray excitation, as in x-ray fluorescence, there is no continuous spectrum generated, and the background consists only of scat-

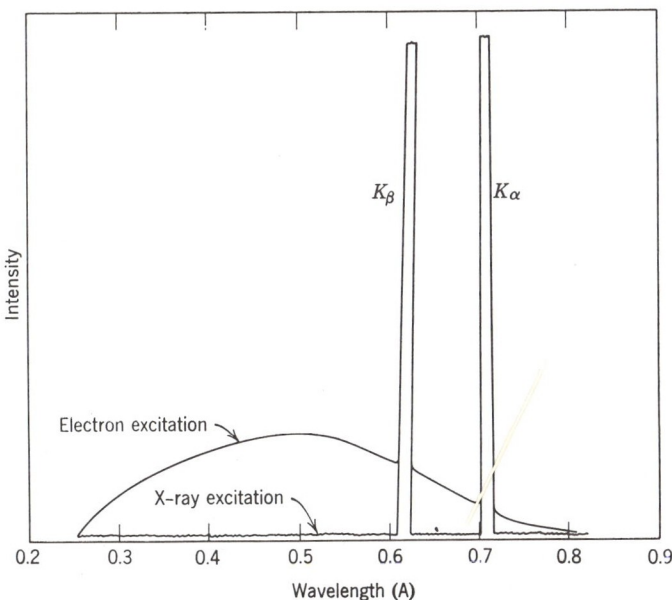

Fig. 1.2 Comparison of molybdenum spectra excited by 35-keV electrons and by x-rays from an x-ray tube operated at 35 keV. With x-ray excitation the background intensity is much lower than with electron excitation.

tered radiation or secondary fluorescence from the crystal or components. Figure 1.2 shows a comparison of typical line and background intensity for a molybdenum specimen excited by electrons and by primary x-rays of the same energy. The line/background ratio is ten times better for x-rays than for electrons at the MoK_α line positions.

At present, the practical limit of detectability in the electron probe is approximately 100 to 500 ppm, although an estimated limit based on statistics alone is of the order of 30 ppm. For comparison, the limit in x-ray fluorescence is about 1 ppm, which again may be extended in special cases to perhaps 0.01 ppm. The limit of detectability for the electron probe should not be interpreted too literally however, because the local concentration of 100 ppm may correspond to an average composition within the total sample of 1 ppm or even 1 part per billion. This is discussed further in Chapter 6.

Another criterion for the limit of detectability is the detectable quantity expressed in grams. From this viewpoint the electron probe surpasses all other instruments (except perhaps the ion probe) because 100 ppm composition in a 10 $(\mu m)^3$ volume corresponds to about 10^{-14} g. The corresponding limit with x-ray fluorescence is about 10^{-8} g; showing that the electron probe is better by 6 orders of magnitude.

1.4 PURPOSE OF THIS BOOK

This book is not intended as a handbook because methods, techniques, and limitations are changing so rapidly that only a loose-leaf handbook could remain current. Rather it is the intent to furnish analysts with general principles and approaches which are needed for proper understanding and use of the instrument and to obtain its maximum capabilities.

It should be recognized that in electron probe analysis, far more than in other analytical methods, the analyst must work very closely with the scientists or engineers who have the materials problem to solve. This is because the interpretations can be grossly misleading if the areas selected are not representative or if interpretation of data is made without understanding the specimen history and likely reactions. Therefore it is hoped that this book will be useful to a wide variety of disciplines as well as to the probe analysts so they will be better able to communicate with each other.

CHAPTER

2

HISTORICAL BACKGROUND

The conception and development of electron probe microanalysis was based primarily on developments in two distinct fields, namely x-ray spectrochemical analysis and electron microscopy. It was instrumentation and techniques in these two fields that combined to make the present electron probe possible.

2.1 X-RAY SPECTROSCOPY

Even before 1920, it was well known that each chemical element emitted characteristic x-ray wavelengths when excited by electrons or primary x-rays of sufficient energy. It was also observed that the intensity of the characteristic lines of an element depended on the amount of that element present in the specimen, as might be expected. During the 1920s and 1930s, von Hevesy (1) and others pursued the applications of x-ray spectrochemical analysis in many chemical problems and also validated the existence of element 72 (hafnium) from its x-ray wavelengths. Thus the stage was set for the development of fluorescent x-ray spectroscopy in the 1940s, when electronic detectors, stable sealed-off x-ray tubes, and large analyzer crystals became available.

Figure 2.1 shows an early plot of the major lines of the characteristic spectra of the elements from atomic number 11 to 92. Of course the elements of lower and higher atomic number may be added to the graph. An advantage of x-ray spectra compared to spectra in the visible or ultraviolet regions is the small number of lines for each element and the orderly progression of wavelengths in each series, according to the relation

$$\lambda \propto \frac{1}{Z^2} \qquad (2.1)$$

where λ is the wavelength and Z is the atomic number.

Another advantage of x-ray spectra is that the lines arise from the removal and replacement of inner-shell electrons in the atom, and these high energy levels are only slightly affected by the physical state or chemical combination of the elements. By deliberately limiting the reso-

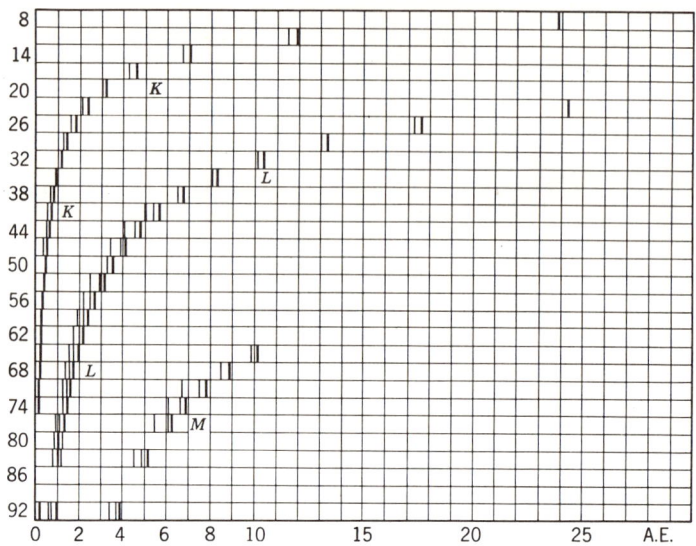

Fig. 2.1 Major x-ray lines for the elements from 8 to 92 from an early x-ray paper.

lution of x-ray spectrometers to about 0.01°, it is possible to resolve all the elements without observing the slight but troublesome shift in wavelength due to changes in valence state which occur especially in elements below atomic number 11.

Figure 2.2 shows a typical modern arrangement for fluorescent x-ray spectroscopy. Primary x-rays from the target of a sealed-off tube, A, operated at about 50 keV and 50 mA, emerge through a large beryllium window and strike an external specimen, B. Intensity is nearly constant over an area as large as 10 cm². Characteristic lines of the elements contained in the specimen emerge in all directions and must be collimated by tubes or blades, C, so that a nearly parallel beam of polychromatic radiation strikes the analyzing crystal, D. At each angular setting, θ, of the crystal, a single wavelength, λ, will be diffracted and pass on to the detector, E, which must be always positioned at angle 2θ to be in proper position to intercept the diffracted radiation. Quantitative analysis is accomplished by measuring the intensity of each characteristic line and relating it to composition.

2.2 ELECTRON MICROSCOPY

During the 1930s and early 1940s, electron microscopy (2) was also developing rapidly, based on research in radio-frequency power supplies,

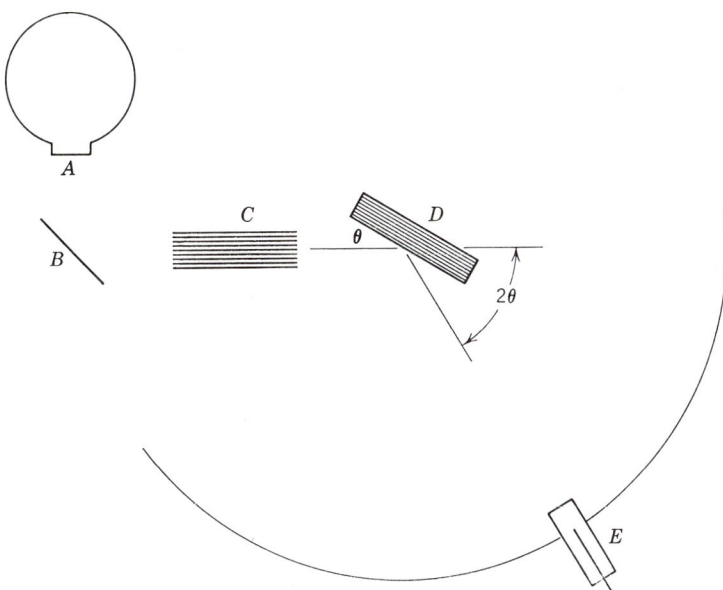

Fig. 2.2 Schematic diagram of the principle of fluorescent x-ray spectroscopy. Key: A, x-ray tube; B, specimen; C, collimator; D, analyzing crystal; and E, detector.

hot-filament electron guns, and the focusing or imaging properties of electron lenses. Von Ardenne (3a), Zworykin et al. (ref. 2b, p. 102) and, more recently, Smith and Oatley (3b), constructed scanning electron microscopes in which a focused beam of electrons was deflected back and forth to cover a square area on the specimen surface. In the refined instrument of Smith and Oatley the backscattered electrons were detected and used to modulate the brightness of a cathode-ray tube, sweeping in synchronism with the electron microscopy beam, in order to give an electron picture of the specimen surface. In other electron microscopes, direct imaging of electrons transmitted through a thin specimen was accomplished with resolution of a few angstroms by utilizing high-quality electromagnetic lenses. Thus the knowledge of electron lens aberrations and corrections developed so that the field was ready for the micron beam-size requirements of the electron probe.

2.3 THE ELECTRON PROBE CONCEPT

The first literature reference that mentioned the use of a focused beam of electrons to excite characteristic x-rays locally and of a simple x-ray spectrograph to measure their wavelength and intensity was a patent application by J. Hillier (4), whose diagram is shown in Figure 2.3. The project was not pursued actively by Hillier however. In 1949, at the Electron Microscope Conference in Delft, Netherlands, Castaing and Guinier (5) presented their first report on actual conversion of an electrostatic electron microscope to focus an electron beam on a solid specimen and an x-ray spectrometer to measure the characteristic x-rays generated (Fig. 2.4). At that time they did not attempt to sweep the electron beam across the specimen as was done in scanning electron microscopes. Subsequently, as reported in his brilliant doctoral thesis in 1951, Castaing (6) improved the instrumentation by changing from electrostatic to electromagnetic electron lenses. He also discussed the relations between x-ray intensity and composition, based on x-ray absorption coefficients and

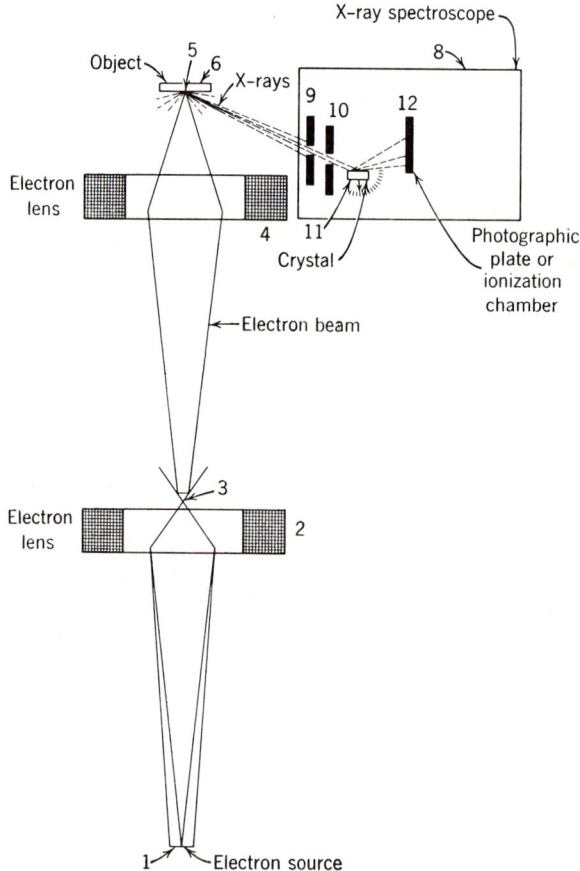

Fig. 2.3 Diagram from Hillier's 1947 patent for an electron probe microanalyzer.

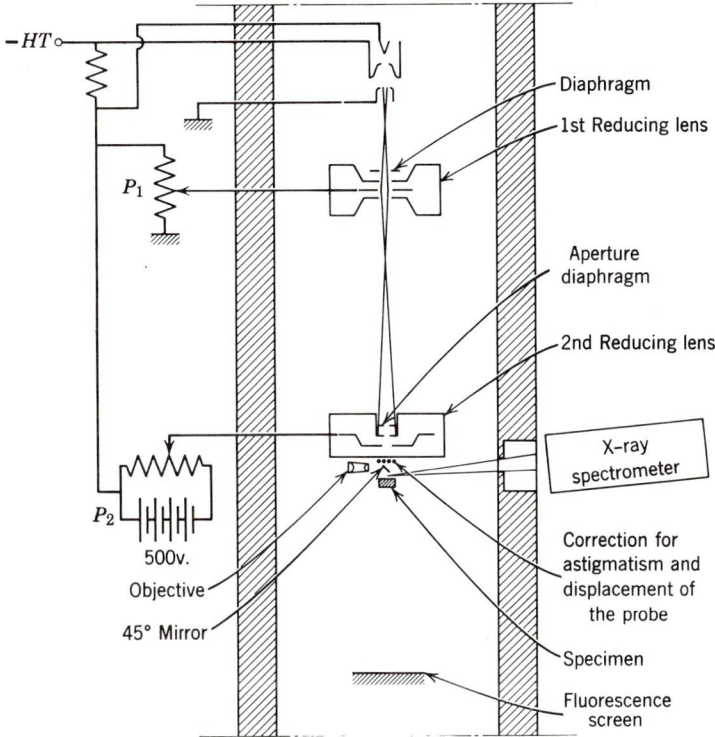

Fig. 2.4 Diagram of the first instrument described by Castaing and Guinier. (Reproduced from report by D. B. Wittry *WAL 142/59-5*.)

the known relations involving x-ray fluorescent yield and fractional absorption by the various electron shells in the atoms. The basic concepts of his thesis have been used by all subsequent workers.

Concurrently with the work of Castaing in France, Borovskii (7) in Russia developed the electron probe concept independently and constructed an instrument of generally similar properties (Fig. 2.5).

2.4 DEVELOPMENT OF A VARIETY OF INSTRUMENTS

During the mid-1950s a number of workers both in Europe and the United States became interested in electron probe microanalysis and began the construction of instruments. They worked independently of each other and along somewhat different approaches. In the discussion below no attempt is made to order the projects chronologically.

Cosslett and Nixon at Cavendish Laboratory, Cambridge, England,

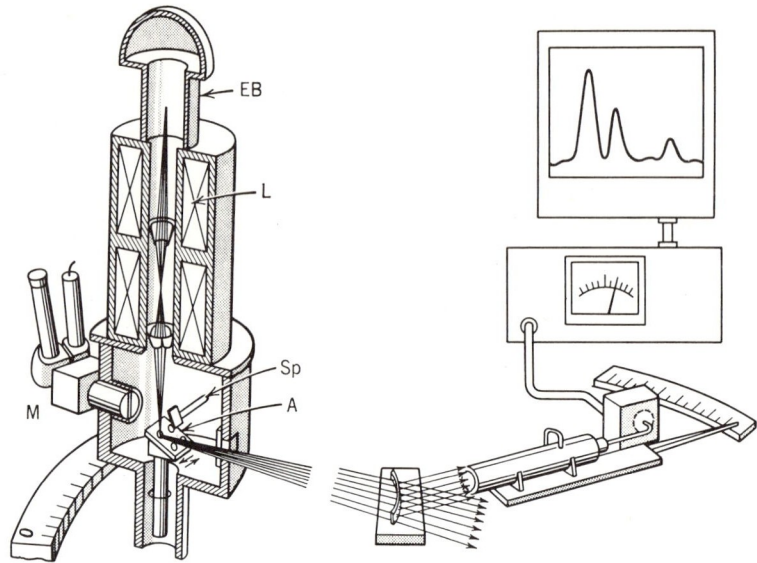

Fig. 2.5 Diagram of the first instrument described by Borovskii (7c)

had developed refined electron optics for an x-ray projection microscope, and it was in their laboratory that they and Duncumb (8) began construction of a scanning electron probe (Fig. 2.6). They swept the focused electron beam back and forth across the specimen in the manner of scanning electron microscopes; there were no visual optics. They used backscattered electron intensity or emitted characteristic x-ray intensity to modulate a television-type display, and obtained both electron and x-ray pictures of the specimen (Fig. 2.7). They did not use wavelength dispersion, as discussed in Chapter 4, but rather energy dispersion as discussed in Chapter 5. Consequently, they were limited initially to high concentrations of elements well-separated in atomic number.

Haine and Mulvey (9) at Associated Electrical Industries, Aldermaston, England, constructed an electron probe (Fig. 2.8) in which the characteristic x-rays were analyzed by a single, scanning, curved-crystal spectrometer similar to those used in x-ray fluorescence. The electron beam was fixed, but the specimen could be translated under the beam in order to measure variations in composition across the specimen. Visual observation and selection of areas for analysis were accomplished by rotating the specimen out of the electron beam to a position in front of an optical microscope. With this instrument Haine and Mulvey were able to measure adjacent atomic number elements in reasonably low

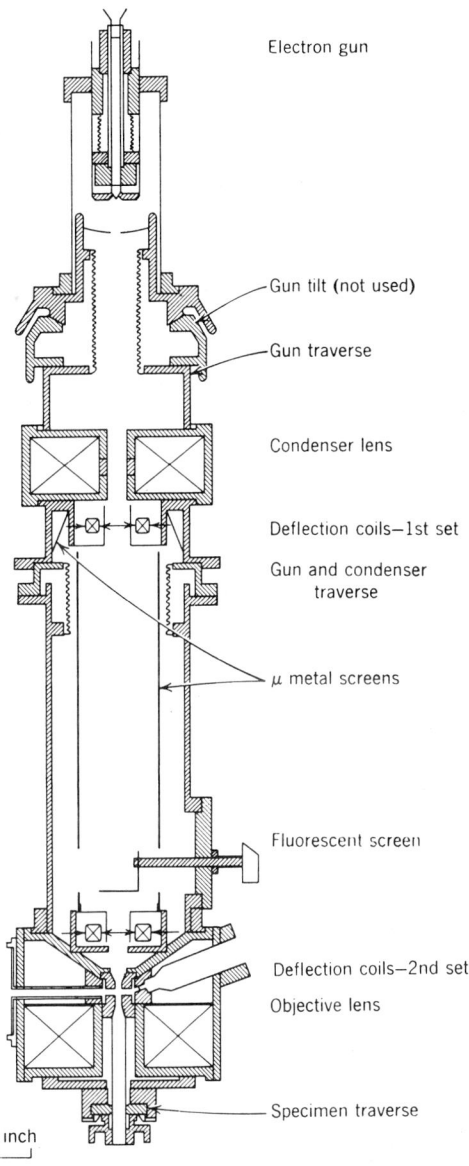

Fig. 2.6 Diagram of the first instrument of Cosslett and Duncumb (8). (Reproduced by courtesy of the authors.)

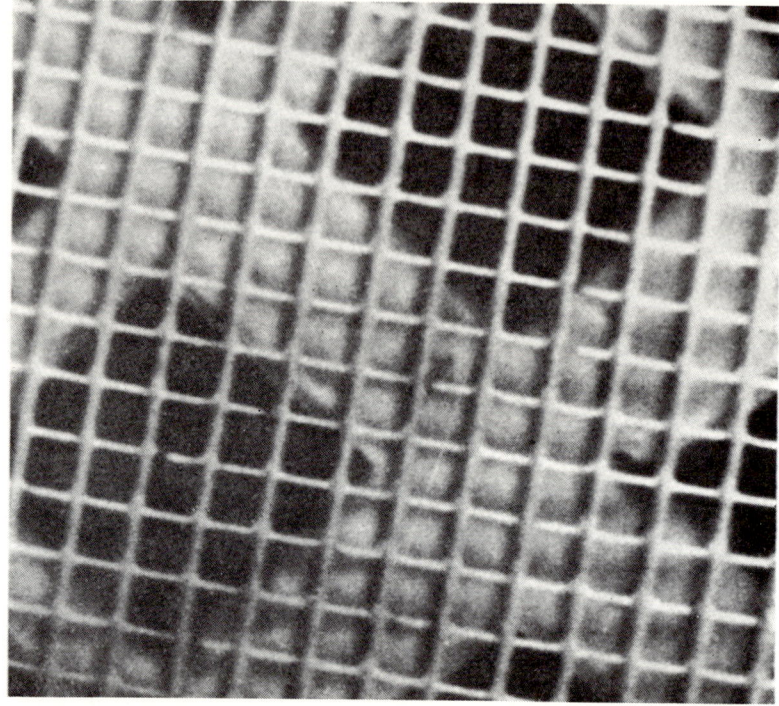

Fig. 2.7 Total x-ray emission picture of two superimposed grids from the early work of Cosslett, Nixon, and Duncumb. (Reproduced by courtesy of the authors.)

concentrations. Birks and Brooks at the U.S. Naval Research Laboratory, Washington, D.C., first experimented with a one-lens electron-optics system of about 5 μm resolution and then constructed at two-lens system (10) of 1 μm resolution (Fig. 2.9). The specimen chamber had a large beryllium window so that curved-crystal x-ray optics systems could be used for simultaneous measurement of 3 to 6 elements (Fig. 2.10). Visual optics were included in the column so that the specimen could be observed during the analysis. The electron beam was fixed and the specimen was translated under the beam by means of orthogonal screws.

Fisher (11) at U.S. Steel Corp., Pittsburgh, Pennsylvania, converted an RCA electron microscope by adding a long working-distance, transfer-electron lens to bring the focused beam and specimen several centimeters out into a large specimen chamber, where the specimen could be viewed conveniently with a conventional microscope during analysis.

Wittry (12) at California Institute of Technology, Pasadena, California, tried to attain a very fine electron source with a field-emission electron gun (Fig. 2.11), but he soon converted to a conventional hot filament.

Schwartz and Austin (13) at Battelle Memorial Institute, Columbus, Ohio, converted an electron microscope and added high-resolution, curved-crystal x-ray optics and a viewing microscope.

Thus in the period from 1951 to 1957, the number and variety of electron probes increased rapidly. By 1956 the first commercial instru-

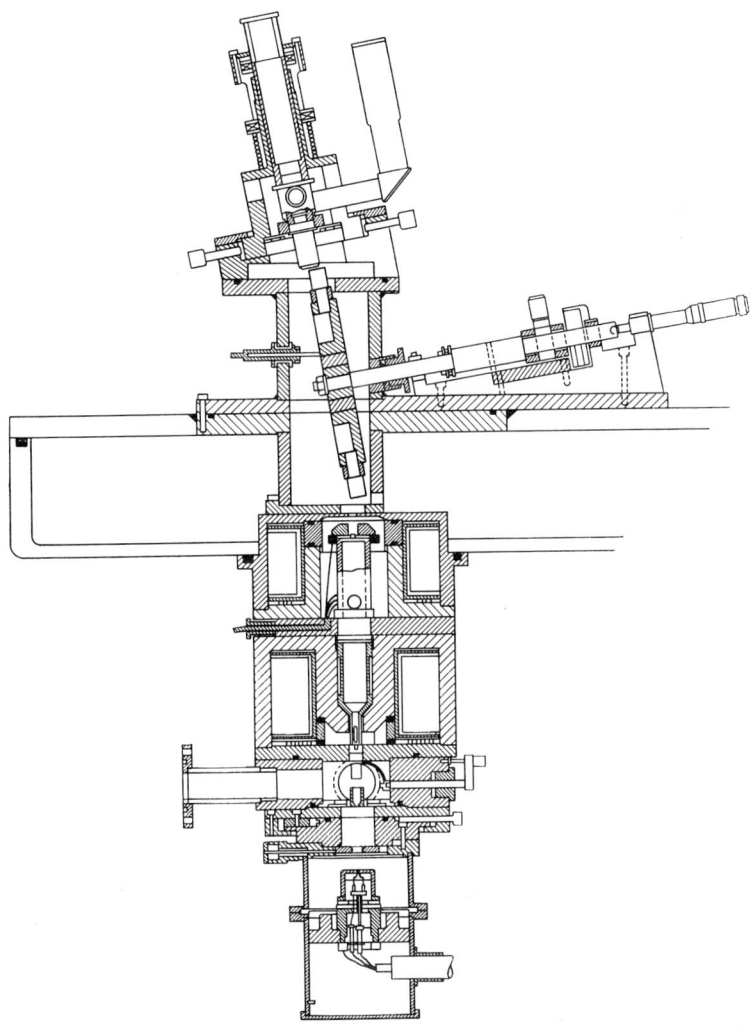

Fig. 2.8 Diagram of the first instrument of Haine and Mulvey (9).

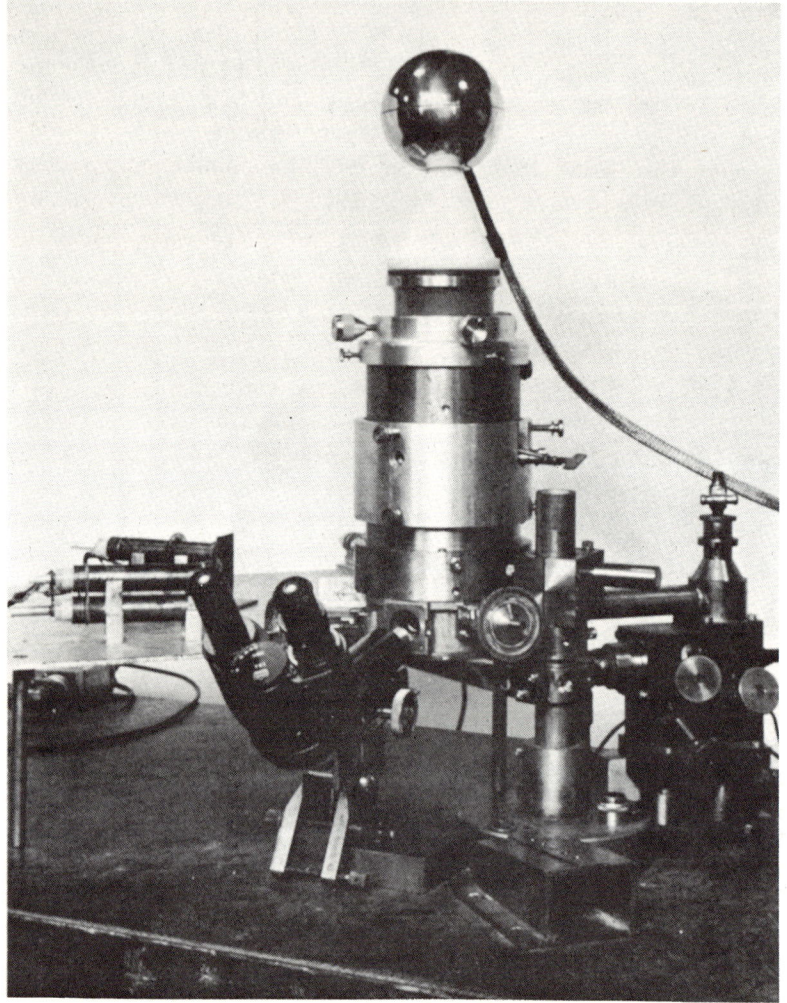

Fig. 2.9 The first electron probe of Birks and Books (10).

ments (Fig. 2.12), built by CAMECA in France, were delivered to ONERA and IRSID in Paris, France. These followed Castaing's design closely.

As with other new techniques, there was a feeling in some circles that although the electron probe was a fascinating instrument, the number of applications was finite, and that once the obvious problems such as identification of precipitates in alloys had been solved there would be very limited need for further instruments. However just as with the electron microscope, as applications of the electron probe began to appear in the literature, it became apparent that the electron probe opened up whole new fields of research in metallurgy, mineralogy, solid-state physics, chemistry, biology, and others. After some hesitation, other commercial interests in electron probe construction developed and at the end of 1969 there were ten known companies* manufacturing instruments; perhaps there are also other commercial manufacturers unknown to the author.

* Europe:
    Associated Electrical Industries, Manchester, England
    Cambridge Instrument Co., Cambridge, England
    CAMECA, Paris, France
    Siemens Halske, Munich, Germany
United States:
    Applied Research Laboratories, Glendale, California
    Materials Analysis Co., Palo Alto, California
    Philips Electronic Instruments, Mount Vernon, New York
Japan:
    Atashi Seisakusho, Tokyo
    Hitachi, Tokyo
    Japan Electron Optics Laboratory, Tokyo

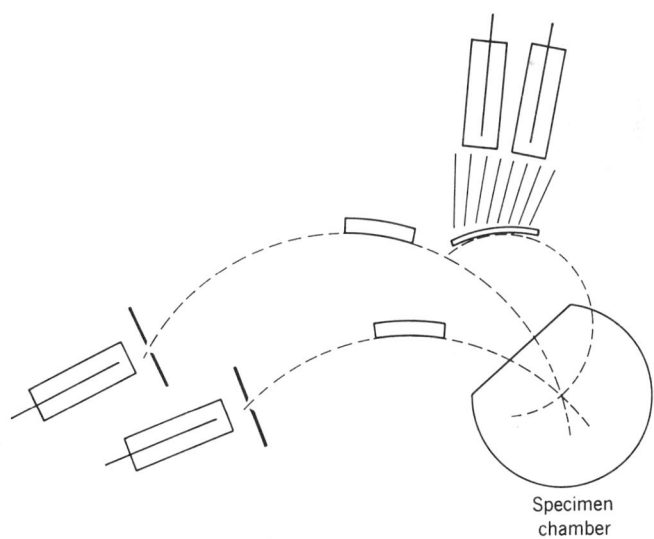

Fig. 2.10 Schematic diagram of reflection and transmission curved-crystal x-ray optics (10).

Fig. 2.11 The first electron probe of Wittry. (Reprinted from his thesis.) (12)

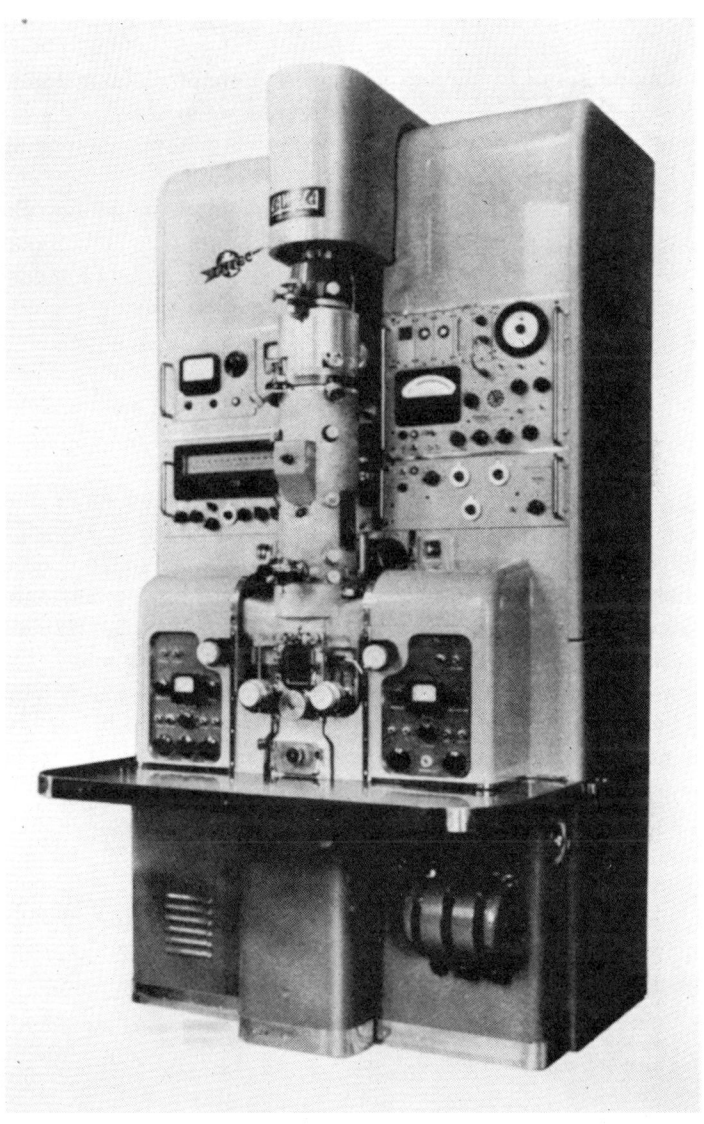

Fig. 2.12 The first commercial electron probe built by CAMECA. [Reprinted from *Laboratoires*, **7** No. 17, (1956).]

2.5 ELECTRON PROBE PHILOSOPHIES

As was pointed out at the beginning of the chapter, the electron probe concept developed from x-ray spectrochemical analysis and electron microscopy. Early workers usually had experience in one or the other of these fields, and their instruments tended to emphasize one or the other concept. For instance, consider the situation in which the worker has specimens whose physical configuration is known from microscopy and it is desirable to know something about the chemical composition of these specimens. This is especially important in solving metallurgical problems already known to exist, such as the identification of precipitates or segregations. Accurate quantitative analysis is not a compelling requirement here—the emphasis is on the qualitative aspect. Several of the early electron probes were adjuncts to microscopes, and the requirement for this type of application is likely to continue.

On the other hand, in the physics or chemistry of solids there are many reactions, such as diffusion or corrosion, in which the change in chemical composition occurs over distances too small to measure by conventional techniques. Both physical and chemical constraints, such as pressure or impurity content, may affect these reactions. By deliberate preparation of specimens to illustrate and delineate these reactions, we can learn much from electron probe analysis. Here the emphasis is on quantitative chemical analysis; microscopy is the adjunct. This viewpoint is of more recent origin than the one in the previous praragraph, but it is expected to gain in prominence.

Still a third viewpoint, as yet only slightly explored, is concerned with composition in the parts-per-billion range. Here the specimen preparation is especially designed to create isolated local concentrations of size and composition that can be measured with an electron probe. The electron probe is thus merely a readout tool for specialized specimen preparation techniques.

2.6 FUTURE OPERATORS OF ELECTRON PROBES

On observation should be made before closing the chapter on historical background. When interest in fluorescent x-ray spectroscopy was revived in the mid-1940s, the technique was adopted almost immediately by analytical chemists. The reason for this was that fluorescent x-ray spectroscopy solved the same problems they had been solving previously by wet chemistry or emission spectroscopy, namely the quantitative chemical analysis of several grams of homogeneous specimen. Specimen-preparation techniques were simple in x-ray fluorescence and the interpretation of data was simpler and more precise than in emission spectroscopy. Chemists found that the x-ray method offered the simplest statistics of any analytical technique and allowed quick evaluation of precision.

During a similar period in the development of electron probe microanalysis, scarcely any analytical chemists became actively engaged in the field. This is probably due to the fact that the electron probe did not solve the same problems they had been solving by other means. Rather it solved problems previouly solved by metallurgists, and thus it was natural that the first enthusiastic users of electron probes should be metallurgists. Now, however, the technique is widely used by chemists, mineralogists, geologists, biologists, and others.

CHAPTER

3

ELECTRON OPTICS COLUMN AND CIRCUITS

3.1 THE VACUUM SYSTEM

Electrons are absorbed or scattered by even a small amount of air; therefore it is necessary to operate the electron optics column including the specimen chamber and electron backscatter detectors in a vacuum of less than 10^{-4} torr. This requires continuous pumping by a diffusion pump. Ion pumping is not practicable because the system is opened frequently to change specimens. Experience has shown that hydrocarbon pump oil (sulfur free) is preferred over silicone oil because even with traps or baffles some of the pump oil molecules drift into the column and are deposited on the specimen by interaction with the electron beam. With hydrocarbon oils the deposit is carbon and causes no interference except in the analysis of very low atomic number (Z) elements. In fact, the deposit is usually helpful in marking the exact location analyzed. Figure 3.1 shows the scan line across the precipitates in a high-temperature alloy. With silicone pump oil the deposit is silicon and it does reduce electron excitation sufficiently to interfere with quantitative analysis.

X-ray absorption, except for elements below sodium, does not require a diffusion pump vacuum in the spectrometers. It may be convenient therefore to have the x-ray spectrometers in air or in a mechanical pump vacuum of 10^{-3}–10^{-1} torr. In this case a window is necessary to separate the x-ray optics from the electron column. The window may be beryllium capable of withstanding differential pressure of an atmosphere for elements above about calcium but must be a thin plastic membrane for elements from sodium to calcium. This membrane will not support atmospheric pressure and must be protected by some kind of closure when either the column or x-ray optics chamber is raised to atmospheric pressure.

The optical viewing system should have the objective lens inside the column, close to the specimen to achieve good resolution. A vacuum window is necessary to bring the eyepiece outside to the observer.

It is convenient to be able to change specimens without breaking vacuum in the main electron optics column. This means an airlock sys-

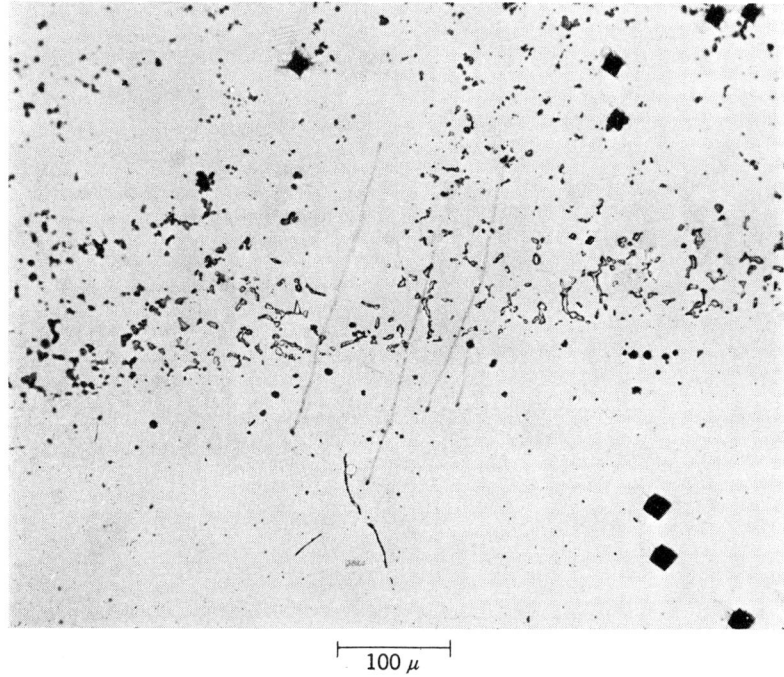

Fig. 3.1 Micrograph of high-temperature alloy showing a sharp line of demarcation of precipitates. The dark squares are diamond-indenter marks used as guides in finding the area of interest. Four dark streaks near the center of the micrograph are the contamination marks left by the electron beam in moving over the region of interest.

tem into which the specimen may be introduced and prepumped before moving it into position in the beam.

3.2 ELECTRON GUN

Most of the present electron probes use a hot-filament electron gun as a source of electrons, as shown schematically in Figure 3.2. The filament wire is about 0.004 in. (100 μm) in diameter, and the effective area at the tip from which electrons are emitted is about 100 × 150 μm. Pointed-tip filaments or cold, field-emission tips are not as suitable as in electron microscopy because of the difficulty in maintaining a very constant electron emission over intervals of half an hour. The filament is biased positive with respect to the cathode shield by the order of 100 to 1000 V to improve stability and to concentrate the emergent

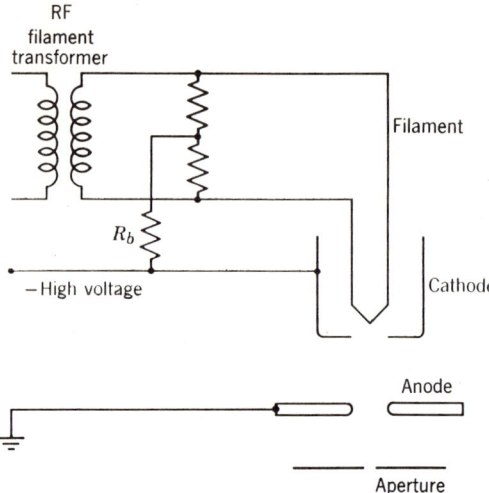

Fig. 3.2 Schematic circuit of the high-voltage and filament supply for an electron gun.

electrons into a nearly parallel beam. The closer the filament tip is to the cathode shield, the greater is the bias voltage required, and the greater is the beam current. The filament temperature must be higher however, and filament life is shorter. Figure 3.2 also shows a common gun-biasing arrangement. The anode and the remainder of the column are grounded for operator protection. The bias resistor, R_B, should be adjustable from about 1 to 10 MΩ during operation to achieve optimum output over the entire high-voltage range. Too low a bias setting will cause a reduction in beam current, and too high a bias may shorten filament life because the filament must be run at higher temperature.

In using the electron gun it is important that the filament be centered with respect to the hole in the cathode shield. It is possible to do this when the filament is installed especially if the filament has been prefired in a separate vacuum system. However it is desirable to have a centering adjustment that can be made when the gun is operating, because often the filament warps when it is hot and moves from its original position. Two other adjustments are also desirable if the gun is to be used over a range of voltage. The first is the adjustment of the filament toward or away from the cathode shield, and the second is the adjustment of cathode-to-anode distance. Because the particular gun parameters determine the optimum settings, it can only be said here that the user must refer to the operating instructions furnished by the manufacturer with each particular gun.

3.3 VOLTAGE AND CURRENT

The effective operating voltage is essentially the difference in potential between the filament tip and the anode rather than between the cathode and the anode. Because of the self-biasing shown in Figure 3.2 it is the cathode voltage which is read on the meter of most electron probes. This introduces an error of a few hundred volts if the meter reading is used in calculations for quantitative analysis (Chapter 7). Usually this will be a negligible difference but may need to be considered if one is working close to the excitation potential for some of the elements in the specimen.

The high-voltage supply should be stabilized to about 1 part in 10^4, over intervals of one-half hour. Such a stringent requirement of say 2 V in 20,000 V is necessary because the beam focusing in an electron optics system is sensitive to the electron energy. A similar stability requirement exists for the current in the electron lenses. If stability in beam focusing is not maintained the beam size will fluctuate by a few microns and errors will result in the interpretation of intensities from small precipitates and narrow diffusion zones.

Based on Langmuir's (14) measurements and calculations, Castaing (15, 16) has shown that the maximum current per square centimeter from a tungsten filament operated at 2700°K in a typical electron gun is limited to about 2 A/cm². For the 100×150 μm source this corresponds to about 150×10^{-6} A. Most of the electrons emitted from the filament do not pass through the anode into the electron optics column however. Thus the beam current from a typical biased electron gun will be only 1 to 10 μA. As stated in the previous section, the beam from the electron gun is nearly parallel; it is also of very small cross-section at the position of the aperture in Figure 3.2, namely about 0.002 in. The purpose of the aperture is primarily to block stray electrons from entering the first lens. This aperture should be of platinum and should be flamed gently in a reducing flame each time the filament is changed in order to prevent contamination buildup with resulting electrostatic charging and beam instability.

3.4 ELECTRON LENS SYSTEMS

To go from the source size of 100×150 μm to the desired beam size of 0.1–1 μm at the specimen requires a total demagnification of 100–1000 times. It is not feasible to achieve this with a single lens and still maintain the desired working distance at the specimen. Therefore one function

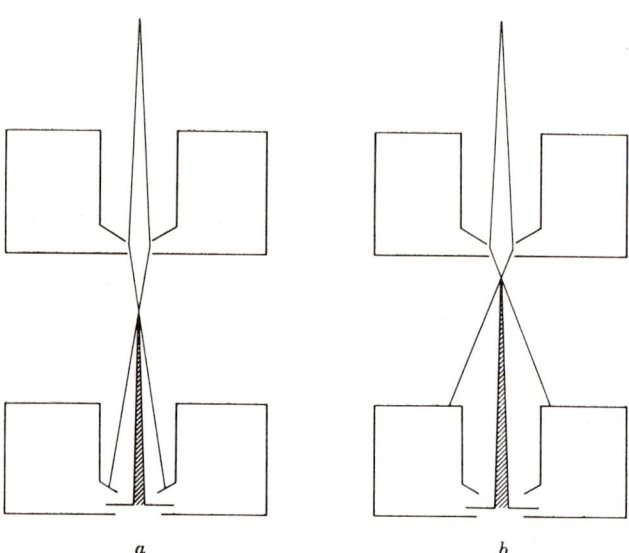

Fig. 3.3 Electron beam intensity at the specimen is controlled by the focusing condition of the first electron lens. In (a) the first lens is focused to allow most of the beam to pass through the aperture of the second lens. In (b) the first lens is set for shorter focal length and only a small portion of the beam passes through the second lens aperture.

of the first lens is to achieve part of the required geometric demagnification by forming an intermediate image. The other function is to control the beam current reaching the specimen. The two functions are interrelated because the greater the demagnification the lower the beam current, as shown in Figure 3.3. In Figure 3.3a there is less demagnification, but a greater fraction of the cone of radiation passes through the second lens aperture. In Figure 3.3b the situation is reversed. Generally the demagnification varies between 10 and 100 times, and the corresponding reduction in beam current passing through the second lens between 100 and 10,000 times. More will be said about the overall demagnification and reduction in beam current in Section 3.7.

Usually the first lens is a strong electron lens with a focal length as short as 4 to 5 mm. Some of the newer commercial instruments use what is called a double-condenser lens as in electron microscopy. That is, two pole pieces are employed to give greater demagnification of the first image as shown in Figure 3.4. The operator can activate just one of the pole pieces if less demagnification is required. For a strong electron lens the aberrations are small, and in electron probe operation the spherical aberration of the first lens usually does not materially affect the final image because the spherical aberration of the second lens is greater. Any astigmatism introduced by the first lens is important however, because it carries on through to the final image. Aberrations are discussed more fully in Section 3.7. Physically, the first lens will have a bore on the order of 3 mm and a gap on the order of 2 mm. The number of ampere turns required will be on the order of 800 for electrons of about 25 keV.

Although the first lenses in the various image-forming electron probes do not differ appreciably in principle or operation except for the double condenser described above, there is wide variation in the design of the second lenses. All of the second-lens designs represent attempts to attain reasonable working distance to the specimen within the limitations imposed by the x-ray and viewing optics and by the lens aberrations, desired beam size, and beam current. Several typical second-lens designs are shown to illustrate the salient features. They are classified somewhat

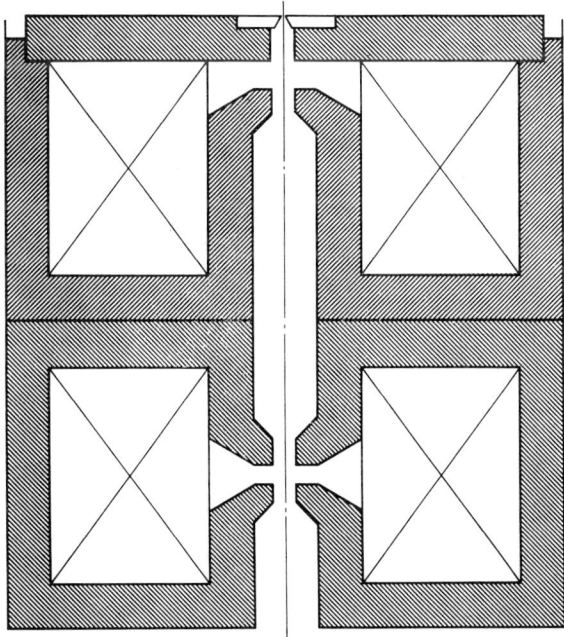

Fig. 3.4 Double-condenser lens used as the first lens in the Philips electron probe By activating either one or both coils greater flexibility in the lens parameters is achieved. (Reprinted by courtesy of Philips Electronics Inc.)

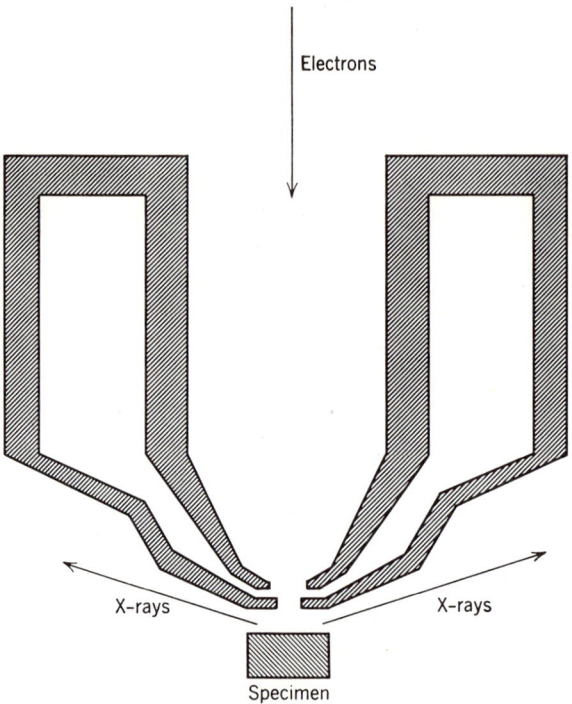

Fig. 3.5 The second-lens configuration of the CAMECA electron probe has an approximately symmetric pole piece.

loosely as symmetric pole piece, asymmetric pole piece, inverted lens, and mini lens.

3.4.1 SYMMETRIC POLE PIECE

The second lens of the CAMECA instrument (15) is shown schematically in Figure 3.5 to illustrate the approximately symmetric-pole-piece design. The bore in both parts of the pole piece is approximately 10 mm. Working distance to the specimen is about 5 mm. The specimen is normal to the incident electrons; emergent x-rays are measured at a take-off angle, ψ, of 18°. The advantage of a large bore in both parts of the pole piece is that viewing optics are easily accommodated in the bore of the lens. A disadvantage, although not serious, is that the magnetic field at the specimen may be as high as 5 gauss. The interaction of this magnetic field with ferromagnetic specimens may deflect the electon beam by several micrometers.

3.4.2 ASYMMETRIC POLE PIECE

Figure 3.6 illustrates the asymmetric-pole-piece design introduced by Mulvey (17) and used in the early Associated Electrical Industries instrument. The bore in the large part is about 50 mm and in the small part is about 5 mm. The advantages are smaller spherical aberration for the same working distance and lower magnetic-field interaction with the specimen. A disadvantage is the difficulty of incorporating in-the-bore visual optics. In the Associated Electrical Industries instrument the specimen is inclined so that incident electrons strike the surface at 30° to the normal. Emergent x-rays are measured in the horizontal plane at a take-off angle of 30°. Other incident and take-off angles could also be used.

3.4.3 INVERTED LENS

Figure 3.7 illustrates the inverted lens design used in the Applied Research Laboratories instrument (18). It differs from the previous designs in that the actual lens coil is below the pole piece, and the specimen is raised into position on an elevated stage. Working distance to the specimen is no longer a critical factor because emergent x-rays are measured back through the pole piece. A variety of visual optics can be accommodated. Electrons strike the specimen at normal incidence, and emergent x-rays are measured at a take-off angle of 52.5°.

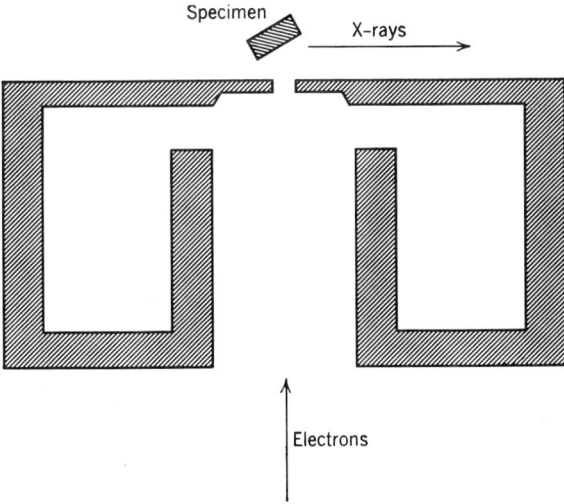

Fig. 3.6 The second-lens configuration of the Associated Electrical Industries electron probe has an asymmetric pole piece with the small bore closest to the specimen.

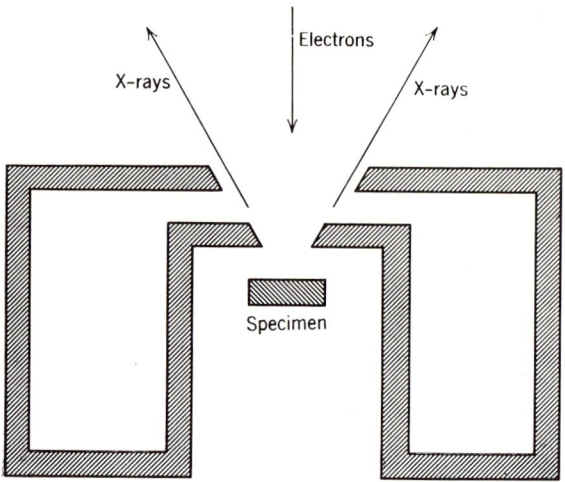

Fig. 3.7 The second-lens configuration of the ARL electron probe allows the x-rays to emerge back through the pole-piece bore.

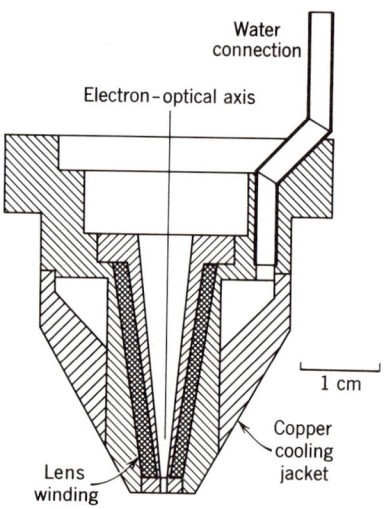

Fig. 3.8 The mini lens as developed by LePoole and redesigned by Duncumb. (Courtesy Tube Investments Ltd.)

3.4.4 MINI LENS

The mini lens illustrated in Figure 3.8 was developed by Le Poole (19). It has no pole piece or yoke and is therefore more like a solenoid coil. Water cooling allows the size to be reduced to under 5 cm diameter. The winding configuration allows the tip of the lens to be close to the specimen but at the same time to employ a large take-off angle. To minimize astigmatism it is necessary that the termination of each winding layer be spaced around the electron-optic axis in a uniform fashion.

3.5 COMBINATION ELECTRON PROBE–ELECTRON MICROSCOPE

Duncumb (20) was the first to build a combination electron microscope–microanalyzer (EMMA) as shown in Figure 3.9. Such an instrument is especially valuable in examining extraction replicas (see Fig.

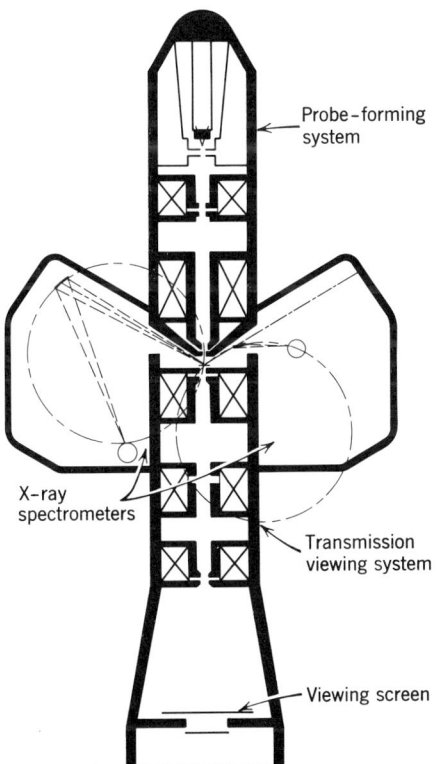

Fig. 3.9 EMMA; a combination electron probe and electron microscope. (Courtesy Tube Investments Ltd.)

6.16), or other thin specimens. The electron microscope allows the size and shape of particles to be measured accurately for absolute intensity determinations. It also allows different kinds of particles to be distinguished quickly so little time is wasted in selecting the areas for electron probe analysis. It also allows electron diffraction capabilities for determination of crystal structure. Early combination instruments did not compare in resolution with standard electron microscopes because the specimen had to be outside the lens field of the microscope objective lens in order to allow egress of the x-ray beam. Use of a mini lens for the final probe lens should remove this restriction and improve microscope resolution considerably.

3.6 BEAM DEFLECTION AND SCANNING

To obtain the display pictures shown in Chapter 6 it is necessary to sweep the beam over the specimen in two orthogonal directions. This

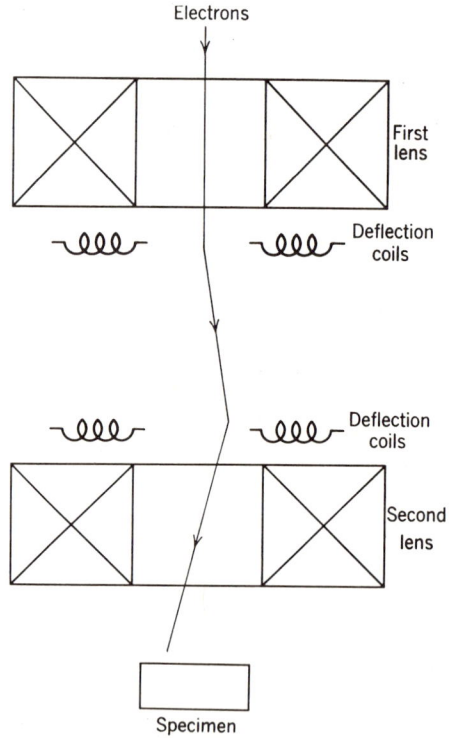

Fig. 3.10 Electron-beam deflection system employed in the Cambridge electron probe.

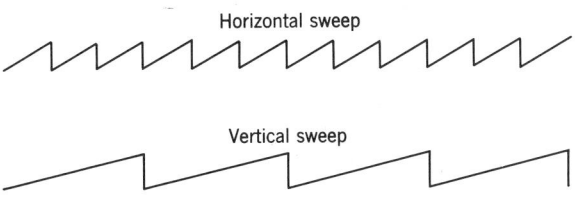

Fig. 3.11 Saw-tooth voltage applied to deflection coils.

is done with deflection coils or plates. Figure 3.10 shows deflection in one direction; the second set of coils or plates is at right angles to those shown. The voltage applied to the coils or plates is in the form of saw-tooth waves as shown in Figure 3.11 resulting in a fast sweep in one direction and a slow sweep in the other direction so as to cover a rectangular area with a set of sweep lines. The frequency and amplitude is variable on both sweeps to control the distance traversed by the beam and the number of lines scanned. Usually either sweep may be turned off to give a single line scan.

It is necessary, of course, to sweep the cathode-ray-tube display in synchronism with the probe-beam deflection but it is desirable to be able to vary the size of the cathode-ray-tube image independently of the probe-beam traverse. This is done by separately variable deflection voltages from the same power supply. The apparent magnification can be changed by reducing the distance traversed on the specimen but keeping the cathode-ray-tube display size constant.

In most electron probes it is possible to sweep the beam over distances from a few microns to a few hundred microns. The larger distances may move the beam so far off the spectrometer focusing circle (see Section 4.9) that the x-ray intensity recorded by the detector is reduced drastically. For this reason some electron probes with high-resolution x-ray crystals use a mechanical translation of the specimen in the direction perpendicular to the focusing circle rather than beam deflection to obtain display pictures. Figure 6.2 in Chapter 6 was taken in the CAMECA Electron Probe with such a mechanical translation.

More examples of displays are given in Chapter 6.

3.7 ELECTRON LENS ABERRATIONS

Electron lens aberrations, rather than theoretical demagnification or electron wavelength, limit the beam size and intensity which may be achieved in electron probes. Electron lenses suffer from all the aberrations known in classical optics and, in some instances, cannot be corrected as well, even theoretically. For instance, spherical aberration can be overcome in classical optics by using a combination of negative and positive lens surfaces or by parabolic lens surfaces. For electron lenses such correction is not possible.

The three aberrations that are most common in electron probes are spherical aberration, astigmatism, and coma.

3.7.1 SPHERICAL ABERRATION

Spherical aberration is the most serious limitation on the minimum beam size which may be obtained for a given beam current. It arises

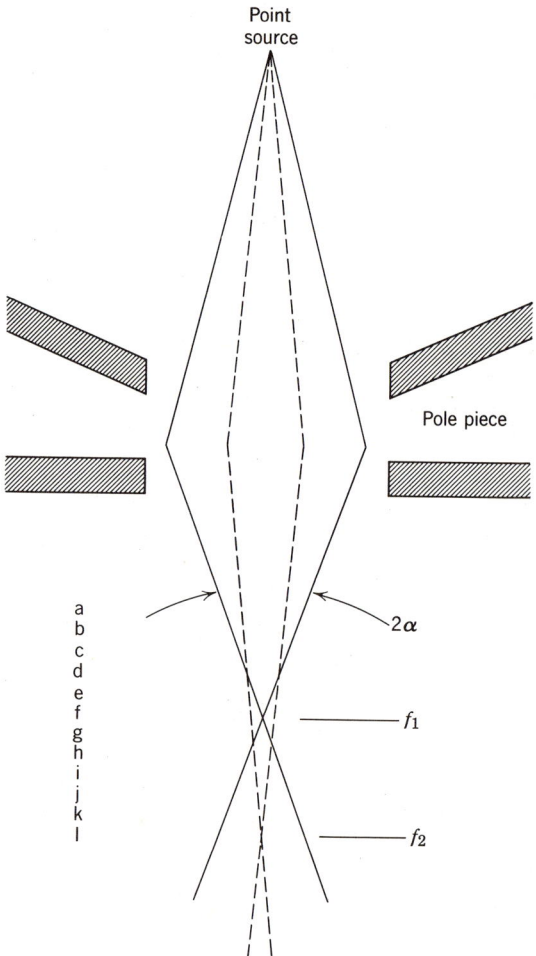

Fig. 3.12 Spherical aberration arises because the focal length for electrons near the axis is longer than for electrons near the periphery of the lens.

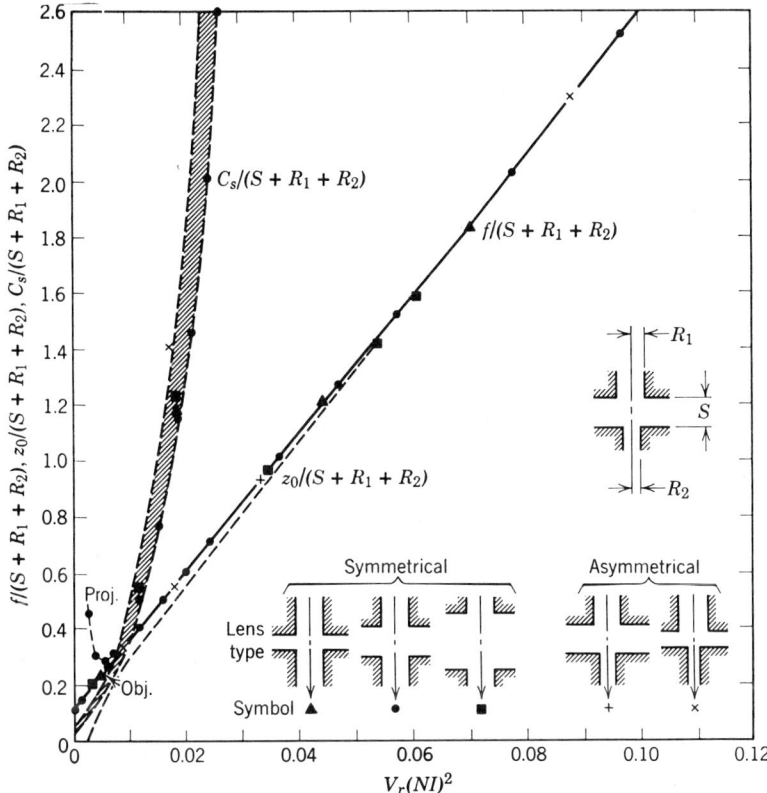

Fig. 3.13 Liebmann's curves relate the spherical aberration constant, C, to the bore, R, and gap, S, of the pole piece for a range of electron energy, V_r, and ampere turns, NI.

because the focal length for electrons passing near the center of the lens aperture is different than the focal length for electrons passing near the periphery of the aperture (Fig. 3.12). In Figure 3.12 α represents the semiconvergence angle of electrons at the image position. The minimum disk of confusion, δ, is given by the equation

$$\delta = \frac{1}{2C_s \alpha^3} \qquad (3.1)$$

where C_s is the spherical aberration constant of the lens in centimeters and depends on the bore size, gap, and average focal length. The term C_s is obtained most easily from the curves published by Liebmann (21) and reproduced as Figure 3.13. The final beam size, D, at the specimen is obtained from δ and the geometric size, d, according to the equation

$$D^2 = \delta^2 + d^2 \tag{3.2}$$

Figure 3.14 shows the appearance of the beam at the positions labeled in Figure 3.12. Positions A, B, C, D, and E represent positions ahead of the crossover and show the hollow ring as would be expected from the ray paths of Figure 3.13. Position F is near focus; G, H, I, J, K, and L are outside the crossover and show a dark central spot surrounded by a lower intensity disk.

Spherical aberration cannot be eliminated completely because it is

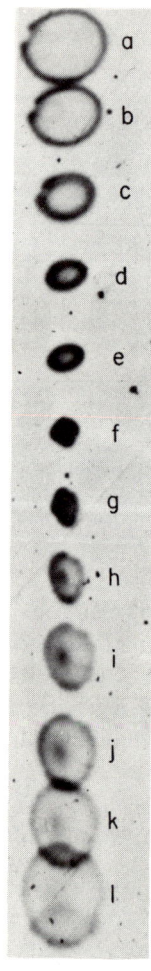

Fig. 3.14 Through-focus series at the positions labeled in Fig. 3.12.

not possible to produce nonspherical electron lenses; neither is it possible to produce a diverging (negative) electromagnetic lens to be used in combination with a converging (positive) lens, as is done with glass lenses for visible light. Spherical aberration may be reduced, however, by placing a limiting aperture in the electron optics system so that electrons that would otherwise pass near the periphery of the lens are stopped. This reduces the convergence angle α in eq. 3.1, and thus the spherical aberration according to α^3. The beam intensity is reduced, of course, by the limiting aperture. For some typical lens parameters spherical aberration is less than 3 cm for beam intensities higher than 10^{-7} A, which is quite satisfactory for most electron probe operations.

The use of nonsymmetric magnetic fields such as quadrapole or octapole lenses also has been suggested by a number of workers as a means of reducing spherical aberration; Castaing (16) has calculated that they might reduce C_s to 0.03 cm, but no practical system based on their use has been developed as yet.

3.7.2 ASTIGMATISM

Astigmatism is caused by a distorted magnetic field in one of the electron lenses. This may be caused by pole pieces which are slightly ellipsoidal rather than circular; it may be caused by the holes in the two halves of the pole piece not being coaxial; or it may be caused by contamination on an aperture which charges electrostatically in a nonuniform manner. The result of astigmatism is a different focal length for the lens in different directions. Figure 3.15 shows a common type of astigmatism for an ellipsoidal lens field. Elongated images are focused at right angles to each other above and below the minimum cross-section. Figure 3.16 shows such astigmatism in a through-focus series for an electron probe with contamination on the second lens aperture. In this case flaming the aperture in hydrogen will remove the contamination and restore symmetric electron optics. However if the astigmatism is caused by ellipsoidal pole pieces or nonconcentric holes in the two halves of the pole piece, it must be corrected by adding a compensating magnetic field distortion to equalize the focal lengths in the two directions. In practice this is done by adjusting soft iron screws around the electron optic axis in the stigmator. Magnetic coils may be used instead of iron screws.

3.7.3 COMA

The third aberration commonly observed in electron probe operation occurs when the electron optics components, such as the electron gun

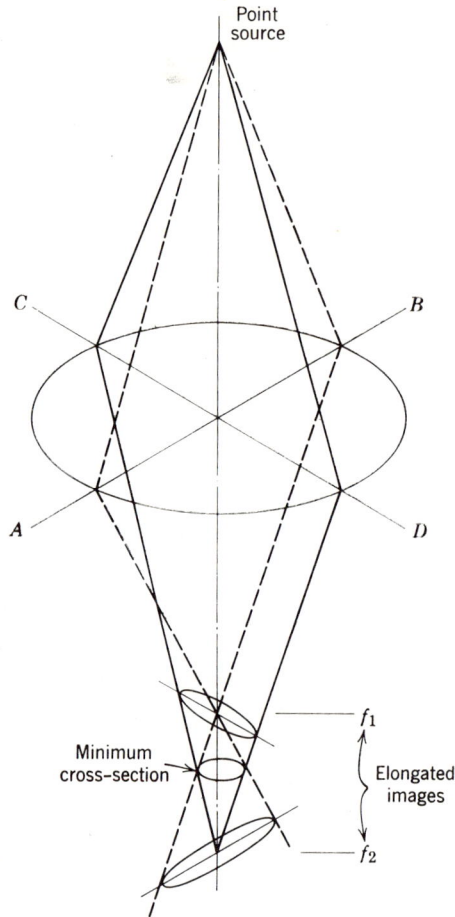

Fig. 3.15 Astigmatism is caused by a distorted magnetic field in the electron lens. The focal length for electron trajectories through A and B is shorter than for electron trajectories through C and D. Elongated images are formed at f_1 and f_2. The minimum cross-section is larger than for a lens free of astigmatism.

and electron lenses, are not aligned properly along the electron optics axis. When either a glass lens (for visible light) or an electron lens tries to image a point off the axis of the lens, the image is elongated, and because of the spiraling action of electromagnetic lenses, is shaped something like the tail of a comet; hence the name "coma." Figure 3.17 shows this effect. Coma occurs because the focal length is not the same for different angles of incidence on the lens. In practice, coma may appear somewhat like astigmatism, because the image is elongated in either case. It is not possible to correct coma properly with the stigmator, although partial compensation may be achieved by distorting the magnetic field differently for the two sides of the lens.

3.8 ALIGNMENT OF ELECTRON OPTICS COLUMN

The purpose of column alignment is to place the electron gun and the two electron lenses on a single axis in order to achieve the best focusing of the beam and the minimum spot size. An example of an alignment sequence is given below. It is designed for one particular electron probe but the concepts are applicable to any two-lens electron probe although the specific steps will be furnished in the manufacturer's operating instructions.

3.8.1 ALIGNING GUN AND FIRST LENS

The image formed by the first lens is the best place to observe this alignment. A fluorescent viewing screen is required just after the first lens and is usually of the removable type that is translated into the

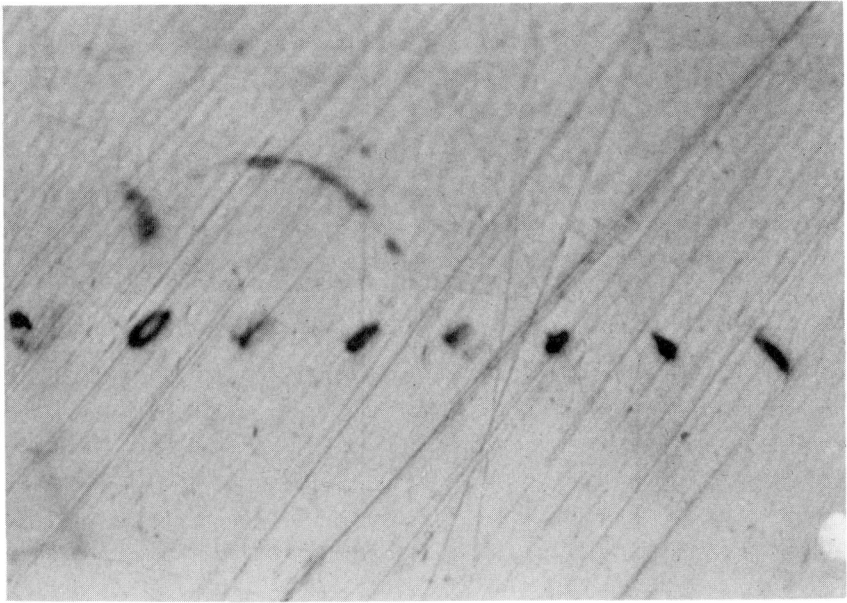

Fig. 3.16 Through focus series with a lens that contains astigmatism. Note that the elongation direction changes by 90° going through focus.

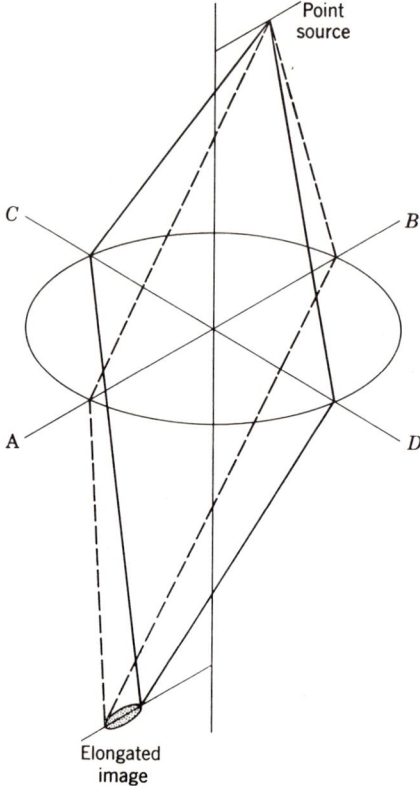

Fig. 3.17 Coma arises because the electron trajectories from a point source off the electron optic axis do not intersect to form a point image.

beam for observation and back out of the way for operation. The lens current is varied in the first lens to underfocus and overfocus the lens during observation. Figure 3.18 shows the appearance of the spot on the intermediate fluorescent screen as the lens current is changed. If the gun and first lens are not in proper alignment the image will spiral, as shown in Figure 3.18a. As the gun (or lens) is translated the spiraling will be decreased, until at proper alignment the image expands about a single position, as shown in Figure 3.18b.

The filament must be centered in the cathode shield or the image will not be as uniform as shown in the illustration. Also, there must not be any tilt of the gun axis with respect to the lens axis, or the image will always move to one side during focusing. Proper manufacture of the components or gun-tilting adjustment is required to eliminate the tilting effect.

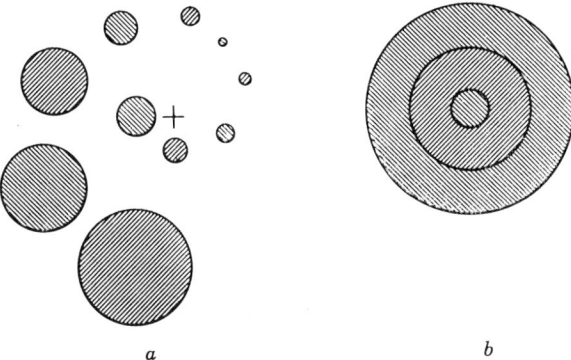

Fig. 3.18 Sequence of intermediate images as seen on a fluorescent screen near the focus of the first electron lens. In (*a*) the electron gun is not aligned with the first lens; as the current is changed in the lens coil, the image spirals as it goes through focus. In (*b*) the electron gun is properly aligned with the first lens and the image remains centered as it goes through focus.

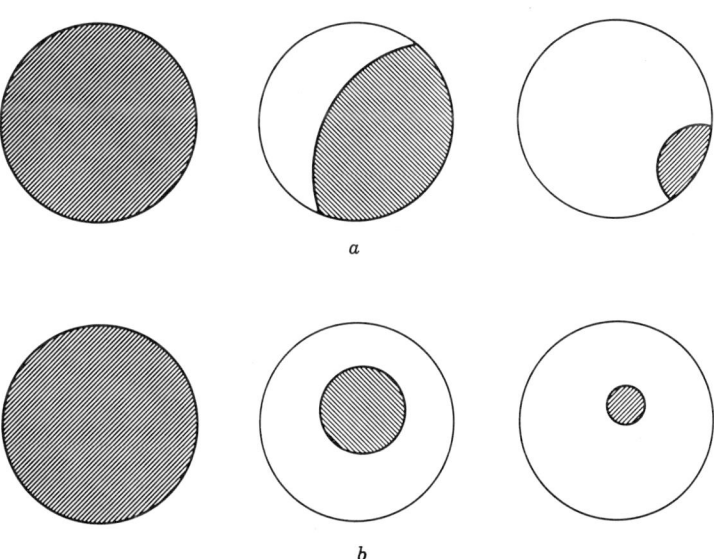

Fig. 3.19 Sequence of images as seen on a fluorescent screen at the specimen position with the second electron lens overfocused to spread the electron beam. In (*a*) the second lens is not properly aligned with the first-lens coil; the image moves out of the second-lens field. In (*b*) the second lens is properly aligned and the image remains within the second-lens field as the current is reduced in the first-lens coil.

3.8.2 ALIGNMENT OF SECOND LENS

Once the gun and first lens are aligned they should move as a unit with respect to the second lens. For alignment of the second lens, observation should be on a fluorescent screen at the specimen position. The second lens is overfocused to give an enlarged spot on the order of 100–200 μm on the fluorescent screen. (The boundary of the spot is actually an image of the field aperture, just as in a light microscope.) Then the current in the first lens is reduced quickly until the illuminated area on the screen moves out of the field of the second lens, as shown in Figure 3.19a. The second lens must then be translated so as to keep the bright spot within the original area of illumination, as shown in Figure 3.19b. The alignment is critical to the extent that the bright spot must remain within the illuminated area if operation is to be satisfactory. It need not be exactly centered however, and usually will not be exactly in the same position each time the high-voltage and the gun-filament current are turned on. Once proper alignment is achieved, it should remain satisfactory for several days of operation but should be checked each time the instrument is turned on if optimum operation is required. In practice, checking both the first lens and second lens alignment requires only about a minute and is very simple for an experienced operator.

CHAPTER

4

X-RAY SPECTROMETERS

Chemical analysis in the electron probe requires separation of the characteristic x-ray lines of the different elements in the specimen and measurement of their intensities. There are two ways to separate the lines: wavelength dispersion with crystal spectrometers, and energy dispersion using the proportional response of electronic detectors. In this chapter we are concerned with crystal spectrometers; Chapter 5 covers energy dispersion.

There are four factors which determine how well the x-ray spectrometer does its job. They are intensity, resolution, line/background ratio, and mechanics. The first three of these are measures of performance and depend on the properties of the analyzing crystal and the detectors and detector slit size. The fourth depends on the design and construction of the spectrometer but influences the other three.

4.1 BRAGG DIFFRACTION

Figure 4.1 illustrates the conditions for diffraction from the planes of a crystal. The increased path length for the ray scattered by the second plane is $2l = 2d \sin \theta$ which must be an integral number of wavelengths, $n\lambda$. Bragg's law states

$$n\lambda = 2d \sin \theta \qquad (4.1)$$

As n takes on the values 1, 2, 3, and so on, wavelengths λ, $\lambda/2$, $\lambda/3$, and so on, satisfy eq. 4.1 and will be diffracted. The first order, $n = 1$, is the principal diffraction and is far stronger than higher orders which are considered as interferences. There are some exceptions to this, mica for example, having many orders of almost equal intensity. More is said about eliminating the higher orders by energy discrimination in Section 5.4. As the angle θ is changed through its complete range from 0 to 90°, the wavelength diffracted will change to satisfy eq. 4.1. Thus the whole spectrum may be diffracted sequentially. The characteristic wavelengths present in the spectrum depend on the elements present in the specimen, and the intensities depend on the relative composition. Unfortunately the relation between intensity and composition is not

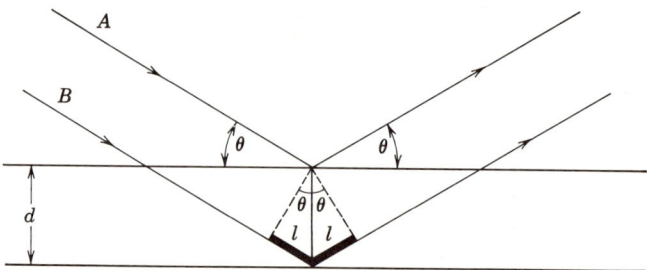

Fig. 4.1 Schematic representation of Bragg diffraction by two planes in a crystal. The spacing is d and the Bragg angle is θ.

linear, and the conversion is discussed in detail in Chapters **7** and **8**. There are two limitations on the wavelength range (assuming one chooses suitable detectors). The shortest wavelength generated, λ_{min}, depends on the electron energy, according to the equation

$$\lambda_{min} = \frac{12{,}400}{V} \tag{4.2}$$

where V is the energy in electron volts. For an operating voltage of 25 kV, λ_{min} is approximately 0.5 Å. Theoretically the longest wavelength, λ_{max}, that may be diffracted depends on the crystal spacing as θ goes to 90° and is

$$\lambda_{max} = 2d \tag{4.3}$$

Practically, because of limitations on 2θ by mechanical features of the spectrometer, λ_{max} is somewhat less than $2d$, usually about $1.8d$.

4.2 CRYSTAL PARAMETERS

The analyzer crystal is the heart of the x-ray spectrometer and its properties help to determine the intensities, resolution, and line/background ratio which can be obtained. Bragg's equation expresses an ideal situation in which all of the atoms in the crystal are arranged in exactly the right positions. Very few real crystals satisfy this condition. Instead, they often consist of small blocks on the order of perhaps 500 Å in size which are nearly perfect within themselves but are misoriented with respect to each other by seconds of arc. These various regions of a crystal will diffract at slightly different orientations of the crystal slab. Figure 4.2 shows the diffracted intensity profile which would result if a real crystal were turned through a beam of strictly monochromatic and parallel radiation. In practice, rocking curves are measured on a double-crystal spectrometer (22) which circumvents the need for strictly parallel radiation.

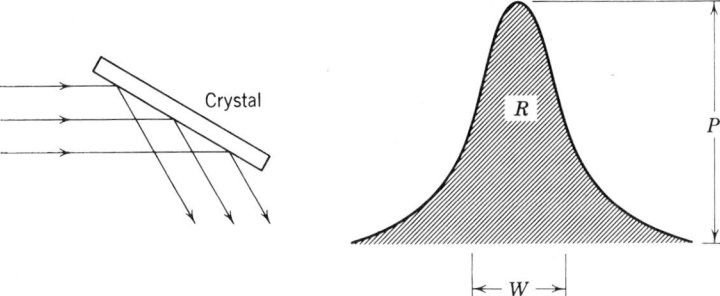

Fig. 4.2 Crystal parameters for diffraction of a parallel, monochromatic beam of x-rays from a single crystal. Key: W is the breadth of the diffraction peak at half maximum, P represents the peak diffraction coefficient when divided by the incident intensity, and R represents the area under the curve and yields the integral reflection coefficient when divided by the incident intensity.

Three parameters are important in Figure 4.2. The term W is the breadth at half maximum; P is called the peak diffraction coefficient and is the fractional intensity diffracted at the peak of the curve, in other words, the ratio of diffracted to incident intensity; R is the area under the curve and is called the integral reflection coefficient. The term R is expressed in angular units, usually radians, because it represents the integral $\int P_\theta \, d\theta$. Table 4.1 gives the W, P, and R values for a few of the common crystals as used in electron probe analysis. These parameters and the crystal optics discussed in Section 4.3 allow us to estimate the expected intensity, resolution, and line/background ratio in Sections 4.5, 4.6, and 4.7.

It is interesting to note in Table 4.1 that the W and R values for LiF are considerably increased by abrading the crystal surface. The

Table 4.1. Parameters of a Few Common Crystals[a]

Crystal	W (sec)	P (%)	R (rad)
Calcite (cleaved and etched)	14	45	4×10^{-5}
LiF (cleaved)	14	40	3×10^{-5}
LiF (abraded)	120	50	4×10^{-4}
Quartz (ground and etched)	20	30	3×10^{-5}
EDDT (sawed and etched)	215	20	2×10^{-4}
KAP (cleaved)	70	13	5×10^{-5}
Graphite (hot pressed)	1800	30	3×10^{-3}

[a] Parameters measured at CuK_α for all crystals except KAP which was measured at AlK_α.

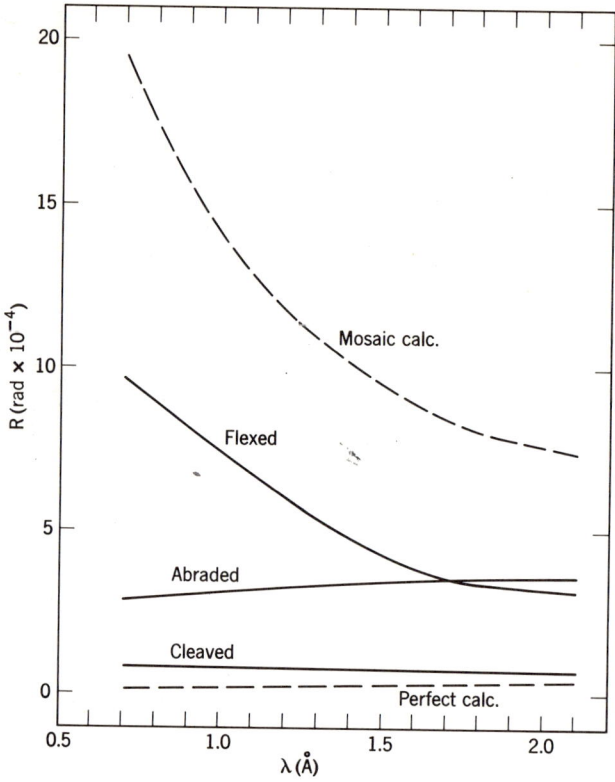

Fig. 4.3 Diffracting power for LiF with different methods of preparation.

explanation generally given is that the edge dislocations introduced increase the angular misorientation while decreasing the primary extinction (23, 24), thus allowing P to remain large and thereby increasing the R value. Flexing LiF such as is done in preparing the curved cyrstals described in Section 4.3 also introduces edge dislocations and gives a similar effect. Figure 4.3 shows R versus λ values for LiF under a variety of conditions (24). The term R is increased more at short wavelengths by flexing than by abrading because the edge dislocations are introduced throughout the volume of the crystal rather than only near the surface. Hailles (25) has shown similar results for abraded silicon. Graphite crystals which are prepared by hot pressing graphite flakes (26) always approximate the nearly mosaic crystal case; the R versus λ curve is similar to the calculated curve for LiF in Figure 4.3, but more than five times higher because of reduced absorption coefficient and a different d spacing.

In other crystals such as calcite, topaz, or quartz, abrading does not introduce dislocations of the type introduced in LiF and does not result in broadening the curve or increasing R if the abrading is followed by etching to remove the disturbed layer at the surface. Neither does curving such crystals result in an increase because the crystals are curved elastically rather than plastically as is done for LiF. Thus curved and ground quartz retains its usual parameters while curved and ground LiF shows the parameters of the flexed and ground crystal of Figure 4.3.

4.3 CURVED-CRYSTAL OPTICS

Because of the point source of x-rays generated in the electron probe, it is desirable to use curved-crystal spectrometers to achieve maximum efficiency. Figure 4.4 shows two possibilities called Johansson (27) and Johann (28) optics after their originators. In Figure 4.4a the crystal is curved to a radius equal to the diameter of the focusing circle and then the surface is ground to fit the radius, r, of the focusing circle. In the plane of the circle, radiation of some wavelength λ emerging from the source, S, will satisfy the Bragg condition everywhere along the crystal, will be diffracted, and will converge to the image, I. As the crystal is moved along the focusing circle, each wavelength will

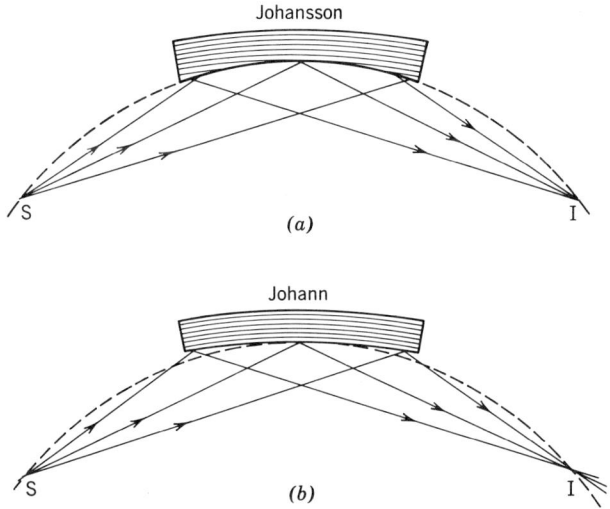

Fig. 4.4 Diffraction from curved crystals. In (a) the crystal is ground to the radius of the circle and is focused to a single image. In (b) the crystal is not ground and the diffracted rays do not cross at a single point.

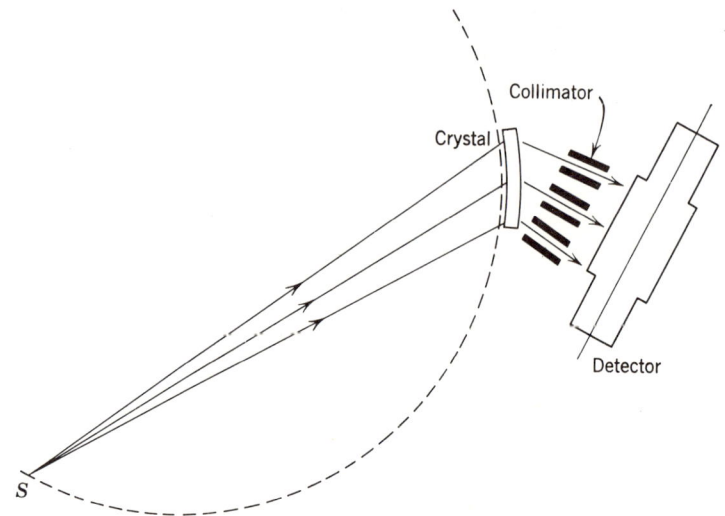

Fig. 4.5 Divergent x-rays of a single wavelength are diffracted by planes perpendicular to the curved transmission crystal. The diffracted x-rays continue to diverge to the detector. A blade collimator prevents scattered or direct radiation from reaching the detector.

be diffracted in turn; the image point will move along the focusing circle at twice the rate of the crystal. In Figure 4.4b the situation is similar except that the diffracted radiation does not converge to a single image point; instead the various diffracted rays are only approximately focused as shown. Johansson optics are preferred for crystals which may be ground because of the better resolution for neighboring wavelengths. Divergence of radiation from the source out of the plane of the circle results in a line image even for a point source; it is discussed in Section 4.4 and illustrated in Figure 4.6.

Curved crystals may be used in transmission (29) as well as in reflection, as shown in Figure 4.5. Again the crystal is curved so that its radius is twice that of the focusing circle, but in this case grinding of the crystal surface is less important. Here the planes *normal* to the surface are used for diffraction. Divergent radiation of the proper wavelength to satisfy Bragg's law is diffracted as shown and continues to diverge. Directly transmitted radiation of other wavelengths must be masked out by the blade collimator so that it does not reach the detector. With a transmission crystal, the solid angle intercepted is increased by the ratio $\cos \theta / \sin \theta$. This is a factor of 5.7 times for LiF crystal and MoK_α radiation. There is a loss of intensity because of crystal absorption however, and the overall increase is on the order of only two to three times for a crystal of optimum thickness.

In some of the early electron probes a so-called semifocusing geometry was used in which the curved crystal was rotated about its axis but remained at a fixed distance from the source. This led to relatively poor resolution except at the wavelength for which the crystal was curved and is no longer extensively used. A discussion of that geometry may be found in the first edition of this book.

4.4 FANNING DIVERGENCE

Divergence out of the plane of the focusing circle is called fanning divergence. Its effect is to increase the breadth of the line; thus it affects

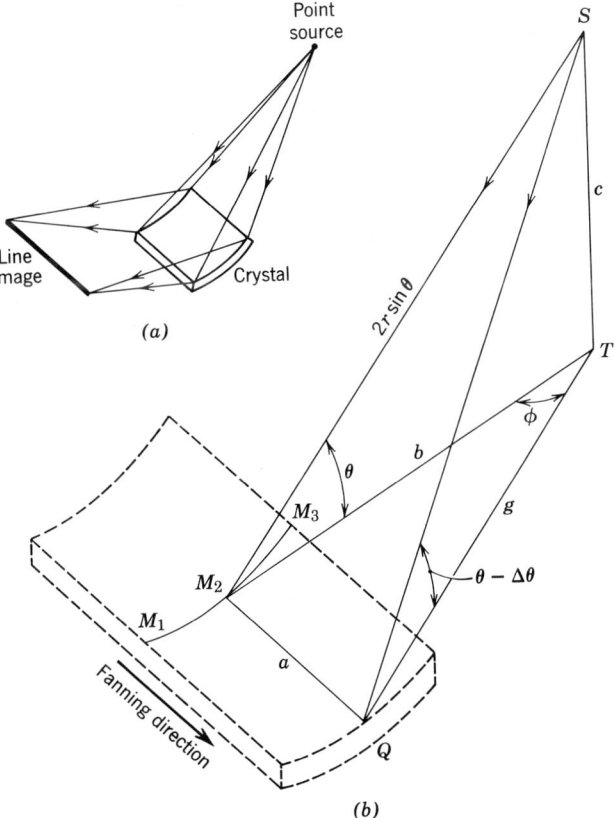

Fig. 4.6 (a) The focusing crystal causes diffracted radiation from a point source to converge to a line image. (b) Details of the fanning divergence and parameters used in eq. 4.4.

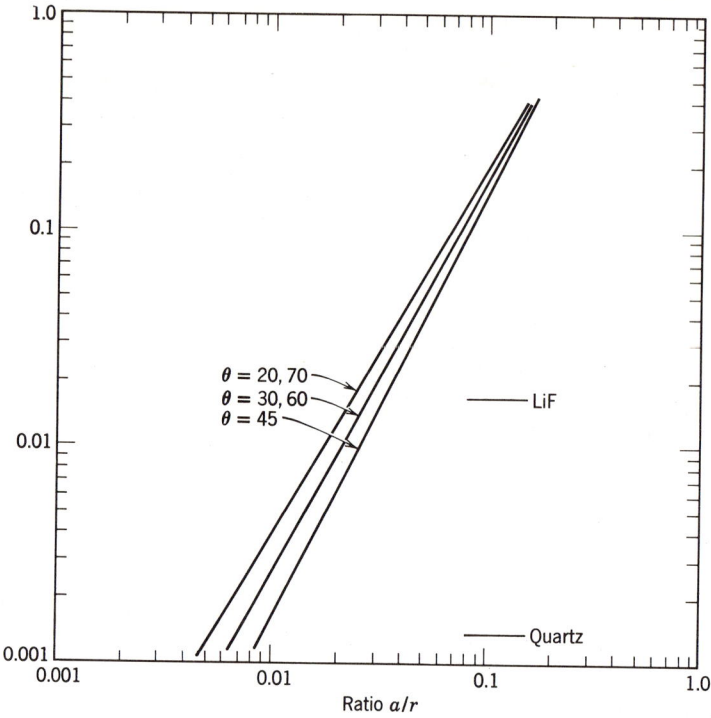

Fig. 4.7 Variation of $\Delta\theta$ against a/r in perpendicular divergence. The separate curves are for different values of θ.

the resolution (Section 4.6), and the limitations on specimen position and scanning size (Section 4.9). Figure 4.6a shows the overall geometry for curved and ground crystals and Figure 4.6b shows the details of the change in θ with fanning angle ϕ. The focusing circle of radius r passes through the points M_1, M_2, and M_3 on the equator of the surface of the curved and ground crystal. The Bragg angle θ will be satisfied everywhere along this equator line from the point source S. Points such as Q, out of the plane of the focusing circle, are associated with the fanning divergence, ϕ. At point Q the angle of incidence with the crystal planes will be $\theta - \Delta\theta$. Let a be the perpendicular distance of Q from line $M_1M_2M_3$. Then $2r \sin \theta$ is the distance from source to crystal; b is the tangent to the crystal at M_2; and c is the perpendicular from S to line b. As θ goes from 0 to 90°, b increases from zero to a maximum value at $\theta = 45°$ and then decreases to zero again at $\theta = 90°$. The term ϕ is the divergence angle and g is the distance from T to Q. The angle between the incident ray and the diffracting planes at point Q is $\theta - \Delta\theta$ and is always less than θ for curved and ground crystals. The relationships are

$$b = 2r \sin \theta \cos \theta = r \sin 2\theta \tag{4.4a}$$

$$c = 2r \sin \theta \sin \theta = 2r \sin^2 \theta \tag{4.4b}$$

$$a/b = \tan \phi \tag{4.4c}$$

$$a/g = \sin \phi \tag{4.4d}$$

$$c/g = \tan (\theta - \Delta\theta) \tag{4.4e}$$

Values of $\Delta\theta$ may be plotted against a/r, as shown in Figure 4.7.

For unground Johann crystals the situation is somewhat different. Figure 4.8 illustrates the situation. The ends of the crystal will be outside of the focusing circle and the radiation from the source S will strike the planes at an angle slightly larger than θ. Let $\Delta\theta'$ be the increase in angle as we move toward the ends of the crystal. If we combine this increase in angle $\Delta\theta'$ with the decrease in angle $\Delta\theta$ as we move out of the plane of the focusing circle there will be a locus of points where $\Delta\theta'$ will just cancel $\Delta\theta$ and the radiation will strike the crystal planes at exactly the Bragg angle θ_0. Ditsman (30) has worked out the geometry and shows that the points satisfying θ_0 describe an X on the crystal surface, as shown in Figure 4.9a, where the slope of the line is given by the equation $\tan \alpha = \cos \theta$. Away from the points that satisfy θ_0 the incidence angle θ is either too small or too large, as shown in the figure. If we allow the deviation from θ_0 to be $W/2$ on each side, then the bands that diffract will be as shown in Figure 4.9b.

Portions of the crystal outside of the bands contribute to the unwanted background and lead to reduced precision in quantitative analysis for a fixed analysis time.

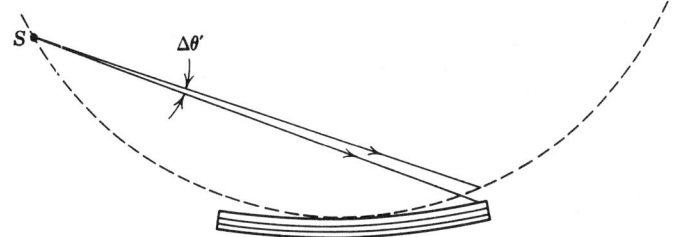

Fig. 4.8 For a curved but unground crystal, the ray striking the end of the crystal will be larger than θ by an amount $\Delta\theta$.

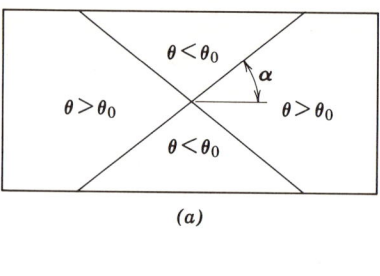

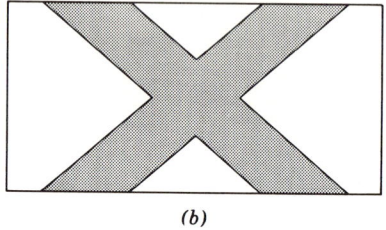

Fig. 4.9 For a curved but unground crystal, the angle for the theoretical locus of points satisfying Bragg diffraction at angle $\theta°$ is in the form of an X. Along the X, decrease in incident angle θ due to perpendicular divergence is just compensated for by the increase due to the crystal surface being off the focusing circle (see Fig. 4.8). The lower figure (b) shows the actual band of diffraction due to the mosaic nature of the crystal and determined by the rocking curve breadth $W/2$ (see Fig. 4.2).

4.5 INTENSITY OF DIFFRACTION LINES

As was stated earlier, the intensity of a characteristic line depends on the crystal parameters and spectrometer geometry. In the electron probe the intensity is always proportional to P because the point source of radiation has a negligible angular spread (note that this is different from the case of an extended source in fluorescent x-ray spectrometry where intensity is always proportional to R). The intensity is also related to electron-beam voltage, beam current, and x-ray yield for each element. Figure 4.10 reproduces total x-ray yield curves for excitation by electrons, protons, and x-ray photons (31). For CuK_α, for instance, and an electron-beam voltage of 25 kV the yield is 2×10^{-4} photons per second per incident electron. Consider a 10-cm-radius focusing circle and a LiF crystal 1.2 cm wide and 5 cm long with P = 30%. The Bragg angle for copper is $\approx 22°\ \theta$. The a/r value for the 1.2 cm wide crystal is $0.6/10 = 0.06$ from Figure 4.7; therefore $\Delta\theta$ due to fanning divergence is $\approx 0.1°$ or 6 minutes of arc. This is considerably greater than the 2 minutes of arc for the rocking curve breadth, W, for LiF (Table 4.1) which means that only a narrow band near the crystal equator will actually diffract within the rocking curve breadth, namely a band about 2 mm wide on each side of the equator. At 22° θ the distance from source to crystal is 7.6 cm and the projected area of the diffracting portion of the crystal as seen from the source is 0.011 sr. At a beam current of 10^{-8}/A the number of electrons per second is 6×10^{10}. Multiplying the several factors together we obtain the expected intensity of

$$(2 \times 10^{-4}) \times (6 \times 10^{10}) \times 0.011 \times 30\% = 4 \times 10^4 \text{ counts/sec}$$

for the pure copper and a 100% efficient detector. This is about the limit of the counting rate capability of electronic detectors and circuits but it will be reduced for lower concentrations or lower currents.

If we had chosen a quartz crystal of 6.7 Å $2d$ spacing the diffracting

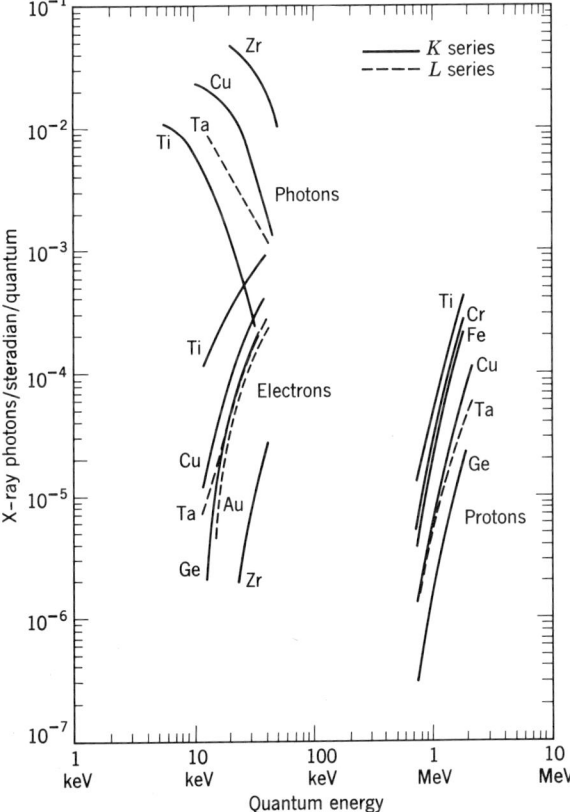

Fig. 4.10 X-ray yield for bulk samples as a function of energy of the exciting quantum.

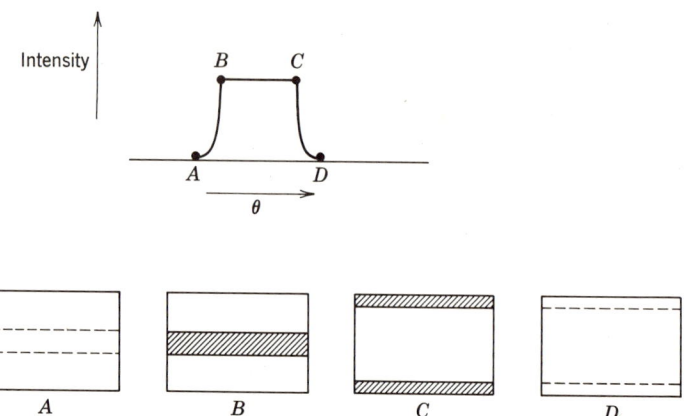

Fig. 4.11 Intensity profile of monochromatic radiation diffracted from a Johansson crystal. Compare the diffracting region with Fig. 4.9 for a Johann crystal.

band along the crystal equation would have been narrower by approximately the ratio W_{quartz}/W_{LiF}, that is, less than 0.4 mm on each side of the equator. The solid angle intercepted by the diffracting portion would be reduced and the count rate lowered to about 1×10^4 counts/sec (taking account of the different θ angle).

4.6 RESOLUTION

The resolution of a curved-crystal spectrometer is not just the rocking curve breadth, W, but is controlled by the fanning divergence which is usually larger. To rationalize this, consider the expected diffraction-peak shape for monochromatic radiation as the crystal is moved along the focusing circle; assume that the detector aperture is long enough in a direction normal to the focusing circle so that diffracted radiation is accepted even at maximum fanning divergence. As the crystal is moved along the circle starting at an angle smaller than the Bragg angle, nothing happens until the tail of the rocking curve begins to be diffracted by the equatorial band on the crystal, position A in Figure 4.11. As the θ angle increases the intensity increases because we approach the peak of the rocking curve for the equatorial band, position B in Figure 4.11. Further increase of θ causes the diffracting band to split with each side gradually moving out towards the edges of the crystal until we reach position C. Then we move down the other side of the rocking curve to position D. Thus the diffracted peak tends to be flattened on top to an extent depending on the width of the crystal and the rocking curve breadth, W. The overall breadth of the diffraction peak will be a convolution of the $\Delta\theta$ spread due to fanning and the breadth W. For the crystal dimensions used in Section 4.5 it would be about 6 minutes of arc for either LiF or quartz. This then determines the resolution for neighboring wavelengths. It is interesting to note that the diffraction line breadth can be reduced and the resolution improved by reducing the detector aperture length in the fanning direction; this causes a reduction in the effective $\Delta\theta$.

4.7 LINE/BACKGROUND RATIO

The line/background ratio, L/B, depends on a number of factors such as concentration of the element, beam voltage, and matrix composition. Limiting values are determined by measuring a pure element and extrapolating to a resolution equal to the natural x-ray line breadth which is about 0.001 Å (32). Table 4.2 shows representative values for practical spectrometers at 30-kV beam voltage and a take-off angle ψ, of 45°. The ratio is better for quartz than for LiF because L/B is roughly proportional to P/R. The reason for this is that the monochromatic intensity is related to P because the negligible source size corresponds to a very narrow angular increment at the peak of the curve but the

Table 4.2. Approximate L/B Ratios at $\psi = 45°$, 30 kV (33)

Line	Quartz	LiF
TiK_α	7×10^3	2×10^3
CuK_α	3×10^3	0.7×10^3
ZrK_α	0.3×10^3	0.06×10^3

Line/Background Ratio Extrapolated to Natural X-ray Line Breadth	
Ti Kα	7150
Cu Kα	2900
Ge Kα	1980
Zr Kα	240
Ta Lα	52
Au Lα	343
Au Lβ	236

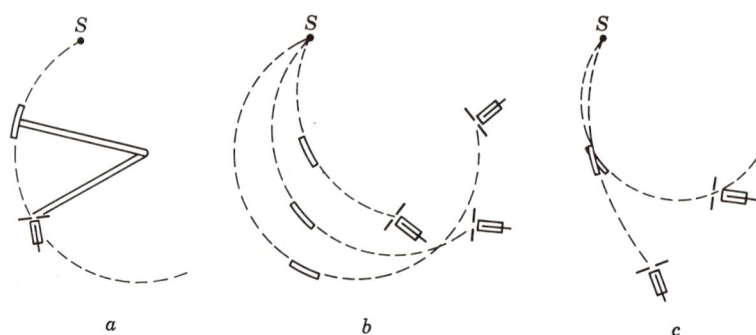

Fig. 4.12 Schematic representation of three arrangements for focusing x-ray spectrometers. In (a) the crystal and detector move on a circle about a fixed center. In (b) the crystal moves along a straight line and rotates about its own center. The detector moves along a noncircular curve to satisfy the focusing-circle requirements. In (c) the distance from source to crystal is fixed and the crystal is flexed to change its radius of curvature. The detector is moved so as to remain on the focusing circle defined by the crystal.

background intensity is related to R because there is some wavelength in the continuum spectrum which will be diffracted by each individual crystallite orientation thus satisfying the whole rocking curve.

4.8 MECHANICS OF SPECTROMETERS

Three types of spectrometer motion are illustrated in Figure 4.12. In Figure 4.12a the crystal and detector move along the arc of a fixed Rowland circle. This is perhaps the easiest system to construct but it is inconvenient because it takes a rather large amount of space and requires a large window for the x-ray beam. Figure 4.12b which is the most popular system shows linear travel of the crystal (swinging the focusing circle) with the detector moving along a noncircular curve so as to be at the proper detecting position. A more compact variation of this is to swing the detector about the crystal at a fixed distance in which case the detector aperture must be larger because the diffracted beam will not have converged to a line image; the larger detector aperture increases the background intensity which is undesirable but can be tolerated in many instances. Figure 4.12c shows a crystal of variable radius of curvature which remains at a fixed distance from the source; the detector also remains at a fixed distance from the crystal. The size of the focusing circle changes with angle θ. The disadvantage of this approach is that very few crystals can be flexed over a sufficient range, the crystal flexing apparatus requires rather high precision, and one does not have the ground surface which gives the best resolution.

Spectrometers may be oriented in either of two planes as shown in Figure 4.13. For the left-hand spectrometer the electron-optic axis lies in the plane of the focusing circle; for the right-hand spectrometer the plane is inclined to the electron-optic axis depending on the take-off angle. The left-hand arrangement allows more spectrometers to be placed around the specimen and is the one usually employed.

Commercial spectrometers often employ two or more crystals in crystal changers which may be adjusted remotely. Generally this does not affect the precision of the spectrometer if the crystals are adjusted in their mounts properly.

4.9 LIMITATIONS ON SPECIMEN POSITION AND SCANNING IMAGE SIZE

Figure 4.14 shows a side view of the specimen and crystal. As the specimen is moved up and down or as the beam is scanned across the surface, the source of x-rays moves off the focusing circle. Suppose we wish to keep the intensity within 10% of the maximum value. This corresponds to 0.39 times the rocking curve breadth, W, neglecting fanning divergence. The vertical height, h, and the scanning size, s, are

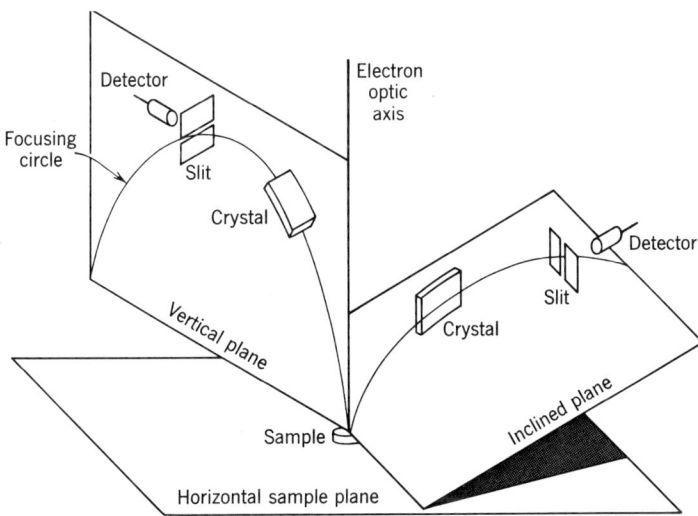

Fig. 4.13 Two ways of arranging curved-crystal spectrometers on an electron probe.

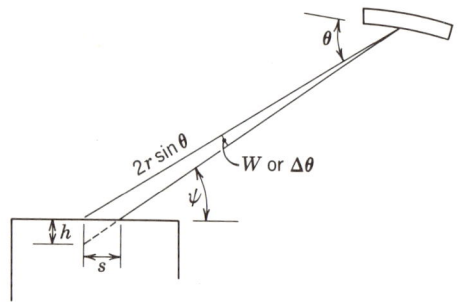

Fig. 4.14 The effective range in elevation of a specimen which falls within the crystal rocking curve acceptance angle.

given by

$$h = 0.39W \times 2r \sin \theta \times \sec \psi \qquad (4.5a)$$

$$s = 0.39W \times 2r \sin \theta \times \sec \psi \qquad (4.5b)$$

For a LiF crystal with $W = 120$ sec, CuK_α radiation, a 10-cm focusing circle and a take-off angle of $\psi = 45°$ where $\sec \psi = \csc \psi$, we find that

$$h = s = 0.39 \times 5.8 \times 10^{-4} \times 20 \times 0.38 \times 1.4 = 24 \ \mu\text{m}.$$

For quartz with $W = 20$ sec the corresponding values are

$$h = s = 0.39 \times 9.7 \times 10^{-5} \times 20 \times 0.23 \times 1.4 = 2.4 \ \mu\text{m}.$$

However the fanning divergence allowed by a crystal 1.2 cm wide will increase these values to more nearly 75 μm for either crystal because we substitute $\Delta\theta$ for W in eq. 4.5. It is interesting to note that the radius, r, of the focusing circle does not affect the calculations when fanning divergence is included because the decrease in the term $\Delta\theta$ is balanced by the increase in the term $2r \sin \theta$.

In practice the 10% variation in intensity allowed in the calculations above are not objectionable at the edges of scanning pictures, but in setting the height of the specimen one would ordinarily try to set to perhaps 1% of the maximum value. This would reduce the multiplier from 0.39 to 0.15 and the h value to less than 30 μm.

CHAPTER

5

DETECTORS AND ENERGY DISPERSION

There are three classes of detectors used on electron probes: x-ray detectors, backscattered electron detectors, and detectors for electroluminescence which may be in the visible, ultraviolet, or infrared regions. These three classes of detectors may be used in a variety of ways to distinguish composition of the specimen; in particular, the x-ray detectors may be used directly for quantitative analysis by energy dispersion as described in Section 5.4.

5.1 X-RAY DETECTORS

X-ray detectors used for electron probe analysis are essentially the same as those used for x-ray fluorescence spectrometers.

PROPORTIONAL COUNTERS

The most common detector is the gas proportional counter in Figure 5.1. The gas will usually be xenon for sealed counters and argon–methane for flow counters. An x-ray photon entering the detector will be absorbed by the gas atoms causing a number of electron-ion pairs to be formed, the number being proportional to the photon energy. Electrons are accelerated toward the central wire while the positive ions are drawn toward the grounded shell. With a voltage of 1000–2000 V on the wire, each initial electron is accelerated and causes many more ionizations so that an avalanche of perhaps 10^4 electrons strikes the wire for each initial electron. All of the avalanches from a single x-ray photon are collected in about 0.1–0.2 μsec; they drop the voltage on the central wire momentarily and this drop is read by the amplifying circuits as a sharp electrical pulse of 0.1–0.2 μsec rise time and 1–2 μsec recovery time. Both the number of initial ion pairs per photon and the number of electrons per avalanche show a statistical spread about the mean value so that for a single photon energy, a pulse amplitude distribution results. That is, a CuK_α x-ray photon has 8000 eV energy and might be expected to cause 8000/12 ion pairs in xenon gas where 12 eV is the ionization potential for xenon. Actually the mean number of ion pairs is more

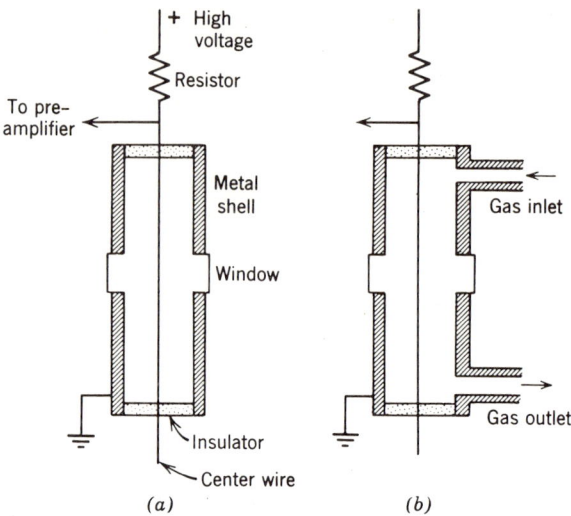

Fig. 5.1 Schematic of (a) sealed proportional counter and (b) flow proportional counter.

like $8000/(2 \times 12) \approx 330$ in xenon-gas proportional counters and there is a statistical spread of $\sigma = \sqrt{8000/(2 \times 12)} = 18$ about the mean. This statistical spread of pulse amplitudes even for monochromatic x-rays leads to the distribution shown in Figure 5.2. A simple estimate for resolution of a gas proportional counter, in percent, is $236/\sigma$ which for CuK_α is $236/18 \approx 13\%$. There are, however, more correct formulae for resolution to be found in the literature (34).

With gas proportional counters the x-ray photons may occasionally ionize a gas atom in the K or L shell and cause it to emit its characteristic x-ray; this leads to an escape peak as shown in Figure 5.3 because the secondary x-ray from the gas atom is likely to escape from the detector without being absorbed. The *difference* in energy between the main peak and the escape peak corresponds in energy to the K or L characteristic x-ray of the gas atom (2957 eV for ArK_α).

Flow proportional counters as shown in Figure 5.1b are used for wavelengths greater than about 3 Å. For such long wavelength x-rays the detector window must be thin so that it does not absorb the incident x-rays before they reach the active volume of the detector. Quarter-mil (6 μm) Mylar is used as the window for x-rays between about 3 and 10 Å but very thin Formar or nitrocellulose (~1000 Å) is required for wavelengths greater than 10 Å. Thin Mylar or nitrocellulose windows are not gas tight; therefore it is necessary to replace the gas in the detector continuously, hence the name flow counter. Mixtures from 90% argon–10% methane up to 100% methane are used for the flow counters.

Generally speaking, counting rates up to 50,000 or 100,000 cps are feasible with gas proportional counters but the higher counting rates often result in a shift of the pulse amplitude distribution as shown in Figure 5.4. This shift is caused by reduced numbers of electrons per

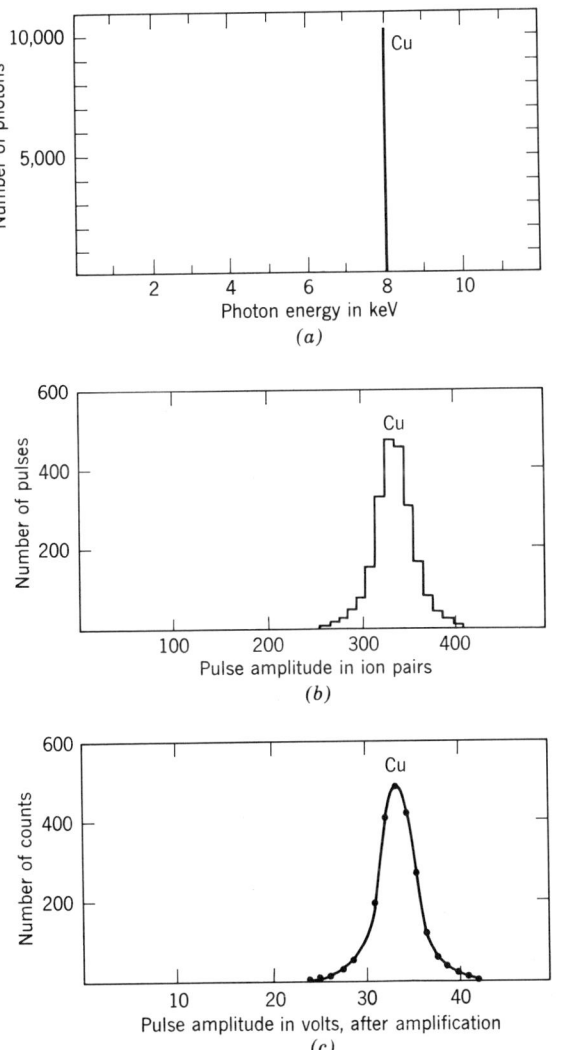

Fig. 5.2 (a) Photon energy in keV; (b) Pulse amplitude in ion pairs; (c) Pulse amplitude in volts, after amplification.

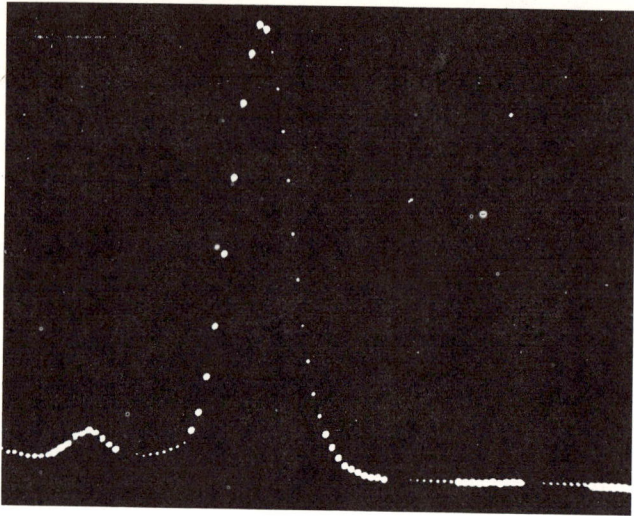

Fig. 5.3 Main pulse amplitude distributions for copper, right, and escape peak, left (argon-methane detector)

avalanche and/or by the positive ion sheath moving out from the wire and effectively reducing the accelerating potential.

When the proportional counter is used with an x-ray spectrometer, as was discussed in Chapter 4, the detector needs only to count the number of x-ray photons per second because the dispersion (selection

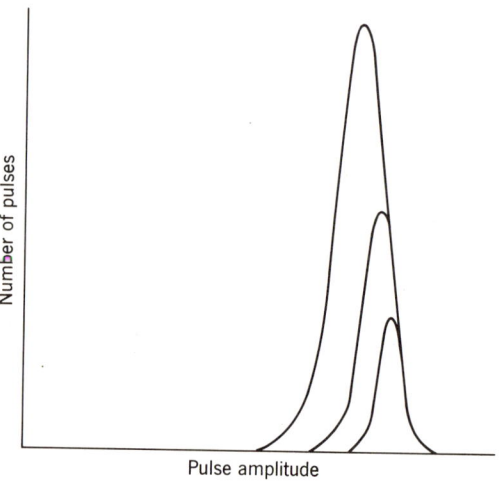

Fig. 5.4 The shift in pulse amplitude distribution with counting rate.

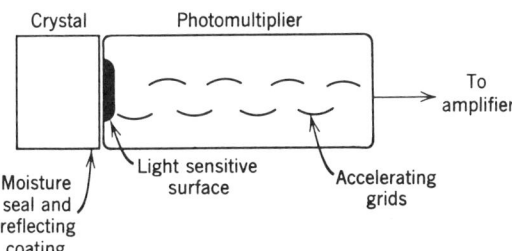

Fig. 5.5 Schematic diagram of scintillator–photomultiplier detector.

of wavelength) is done by the crystal. When the detector is used without a spectrometer as in Figure 1.1 and accepts the whole spectral range of wavelengths from the specimen, the separation of the different characteristic lines must be done electrically as discussed in Section 5.4. This mode of operation is called energy dispersion or nondispersive geometry.

SCINTILLATION DETECTORS

Scintillation-photomultiplier detectors as shown in Figure 5.5 have better efficiency than xenon-gas proportional counters for x-ray wavelengths shorter than 1.5 Å. The scintillator is usually thallium-activated NaI which must be coated with an impermeable membrane because it is hygroscopic. When an x-ray photon is absorbed in the crystal it generates a number of visible-light photons; the number is proportional

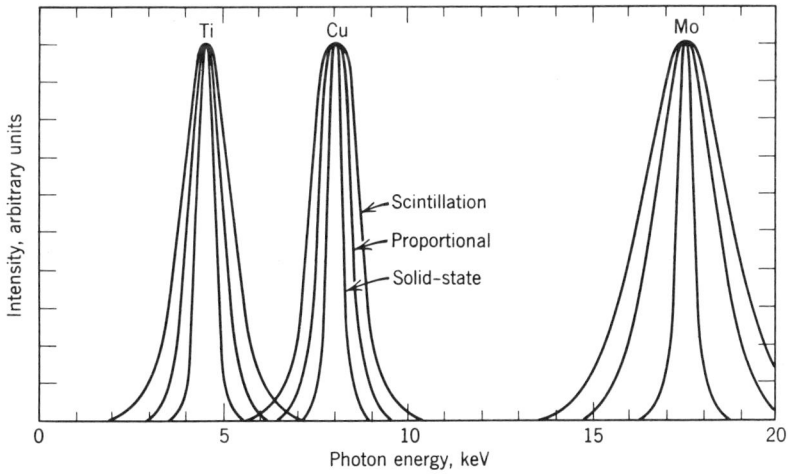

Fig. 5.6 Expected resolution of the three common detectors at titanium, copper, and molybdenum wavelengths.

to the x-ray photon energy. A fraction of these visible photons are detected by the photomultiplier tube and result in an electrical pulse. Again as in the gas proportional counter, there is a pulse amplitude distribution for monochromatic incident x-rays. For the scintillation counter the distribution is two or three times broader than for a gas proportional counter as shown in Figure 5.6.

Occasionally a combination or "piggy-back" arrangement is used with a flow proportional counter in front of a scintillator counter. The longer wavelength x-rays are measured by the proportional counter while the shorter wavelengths pass through and are absorbed in the scintillator.

SOLID-STATE DETECTORS

The newest type of x-ray detector is single-crystal semiconductor material such as silicon or germanium. Figure 5.7 shows the detector schematically. The silicon has low concentrations of other elements added in layers to make the surface p-type (positive), the middle layer i-type (intrinsic or neutral), and the back layer n-type (negative). When a voltage of 300–900 V is applied, no current should flow in the detector. If an x-ray photon is absorbed in the middle layer, it generates a number of electron, positive-hole pairs. The electrons and holes have mobility in the silicon and are quickly drawn to the front and back surfaces respectively resulting in an electrical pulse. The average energy required to generate an electron–hole pair is about 3.8 eV for silicon corresponding to a large number of events per x-ray photon and resulting in a small value for the relative statistical spread of the pulse amplitude distribution. This means better resolution than either the gas proportional or scintillation counters. The theoretical resolution for silicon detectors (<200 eV for MnK_α) was limited originally by electronic and thermal "noise" but is being achieved in newer commercial equipment. In practice the resolution depends on the volume and shape of the detector and

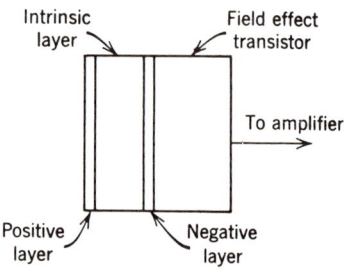

Fig. 5.7 Schematic diagram of solid-state detector and attached field-effect transistor.

on the bonding to the amplifier stages. Even a resolution of 400 eV is only achieved by cryogenic cooling; at room temperature the statistical spread would correspond to x-ray photon energies of several keV and would mean extremely broad pulse amplitude distributions.

Operation at cryogenic temperatures usually means vacuum operation in order to prevent condensation of moisture. In a vacuum spectrometer at cryogenic temperature the solid-state detector can be used to advantage for wavelengths as long as 5–6 Å before the 200 eV resolution is poorer than that of gas proportional counters.

DEAD-TIME CORRECTIONS

Dead time, t_D, depends on the detector recovery rate and on the electronic circuitry response rate; it will vary from detector to detector and from one circuit to another but generally should not exceed a value of about 3 μsec. Once the dead time is known for a particular detector and electronics system the true counting rate, R_{true}, is determined from the measured counting rate, R_{meas}, by the equation

$$R_{true} = \frac{R_{meas}}{1 - t_D R_{meas}}$$

Dead time for the electronics can be determined independently by replacing the detector with a double pulser and noting the closest pulse separation which can just be resolved. For the detector it can be approximately measured with a triggered oscilloscope by noting the time after the initial pulse when pulses again appear. The best overall estimate of dead time for the combination of detector and circuitry is obtained by actually measuring the counting rate loss at high counting rate. Formerly this was done by inserting metal foils ahead of the detector to act as absorbers and measuring the deviation from linearity of the plot of log intensity versus number of foils. With enclosed spectrometers it is no longer convenient to insert and remove foils in the x-ray path so another method of determining dead time must be employed. The one described here is generally similar to the method described by Heinrich (35).

Choose a pair of lines of known intensity ratio, for instance the K_α and K_β lines of a single element. Carry out the following procedure.

1. Adjust the beam current so that the counting rate of the K_α line is about 1000 Hz; the counting rate at the K_β line will then be about 150 Hz. There should be negligible counting losses at these rates.

2. Determine the actual counting rate, R_α, and R_β. In order to achieve reasonable precision in R_α and R_β it is necessary to count for a sufficient fixed-time interval to obtain 10,000–20,000 total counts at the β-line

position (this same time interval will be more than sufficient to give reasonable statistics for the α line).

3. Form the ratio R_α/R_β to be used in step 6.
4. Increase the beam current to raise the counting rate at the β line to about 1000 Hz.
5. Determine the new counting rates $R'_{\alpha\,\text{meas}}$ and $R'_{\beta\,\text{meas}}$. $R'_{\beta\,\text{true}}$ will be essentially equal to $R'_{\beta\,\text{meas}}$ at the 1000 Hz rate but $R'_{\beta\,\text{true}}$ will be greater than $R'_{\alpha\,\text{meas}}$ because of counting losses at the nearly 10,000-Hz rate.
6. Find the true counting rate $R'_{\alpha\,\text{true}}$ from the relation

$$R'_{\alpha\,\text{true}} = \frac{R_\alpha R'_\beta}{R_\beta} \qquad (5.1)$$

7. Find the dead time, t_D from the relation

$$R'_{\alpha\,\text{true}} = \frac{R'_{\alpha\,\text{meas}}}{1 - t_D R'_{\alpha\,\text{meas}}} \qquad (5.2)$$

Note that in the above procedure the background has been neglected under the assumption that the relative contribution of the background remains a constant fraction as the beam current is changed.

As a numerical example, suppose $R_\alpha = 1000$ Hz, $R_\beta = 150$ Hz, and $R'_\beta = 1500$ Hz. Then $R'_{\alpha\,\text{true}} = (1000/150) \times 1500 = 10,000$ Hz. But suppose we measure $R'_{\alpha\,\text{meas}} = 9700$ Hz. From step 7 we calculate t_D from the equation

$$10,000 = \frac{9700}{1 - 9700 t_D}$$

$$t_D = \frac{30 \times 10^{-6}}{9.7} \approx 3 \ \mu\text{sec}$$

Dead-time corrections should be made to all counting rates over 1000 Hz if the numbers are to be used in quantitative analysis.

EFFICIENCY AND RANGE OF USEFULNESS

Efficiency of the various detectors is shown in Figure 5.8. Table 5.1 gives the useful wavelength range for each type of detector. The scintillator is always poorer in resolution than the other detectors but does have better efficiency for the shorter wavelengths.

5.2 ELECTRON-DETECTORS

Backscattered electron detectors are either scintillator–photomultipliers or surface-barrier silicon solid-state detectors (36). Activated

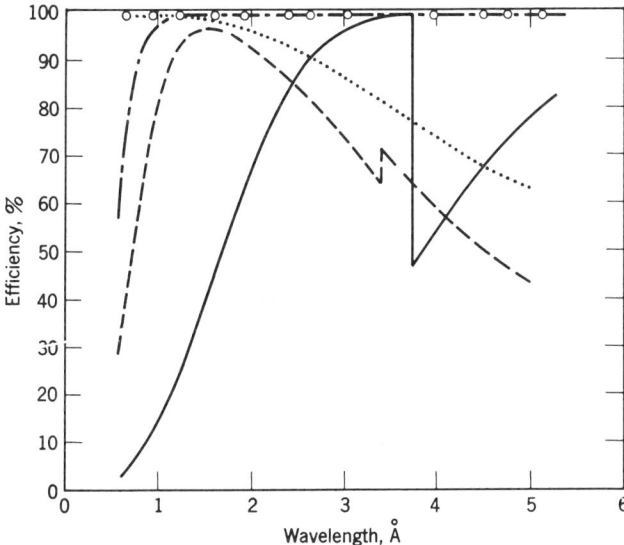

Fig. 5.8 Efficiency for various detectors as a function of wavelength. ———, Ar flow proportional (¼ mil Mylar); – – –, Xe-sealed proportional 1 mg/cm² mica); . . . ., NaI scintillation (1 mil Mylar); – – – – , solid-state silicon. O, Solid-state germanium.

CaF$_2$ rather than NaI is often used as the scintillator because it is not hygroscopic and therefore does not need an impermeable membrane cover which would absorb the electrons. The surface-barrier solid-state electron detectors are used in the current rather than the pulse mode which means they do not require cryogenic cooling but the sensitivity is reduced. Backscattered electron detectors are used primarily for qualitative displays as discussed in Chapter 6. However the backscattered fraction is a function of atomic number (see Fig. 7.5) and occasionally

Table 5.1. Useful Wavelength Range for Detectors

Detector	Range (Å)
Sealed xenon proportional counter	0.7–5
Argon–methane flow counter	2–20
Methane ultrathin-window counter	20–60
Scintillation counter	0.1–5
Silicon solid-state detector	0.3–5[a]

[a] 5-mm-thick silicon with 0.005-in. thick beryllium window.

analysts have used electron intensity as a direct measure of composition in binary diffusion couples (37).

Specimen current (collected electrons) is measured simply by isolating the specimen from ground potential through a 1–10 MΩ resistor and micromicroammeter. The micromicroammeter reading may be used for scanning displays just as the backscattered current is used.

Because the electron current, either backscattered or collected, comes only from the specimen area irradiated by the electron beam ($0.1\text{--}1\mu$), electron display pictures usually show better resolution than x-ray display pictures which represent the larger effective x-ray source size ($1\text{--}5\mu$).

5.3 LUMINESCENCE DETECTORS

Photomultiplier tubes with sensitivities in the visible, ultraviolet, or infrared are used to measure the electroluminescence which is often quite brilliant, especially from mineral specimens. These detectors may be mounted on the viewing microscope or may look directly at the specimen. The output may be used in scanning displays and may be quite sensitive to composition if optical filters are used to discriminate in favor of the desired wavelengths (see Fig. 6.3). Relationships between color of luminescence and composition are more complex than for the x-ray or electron spectra and the results are only semiquantitative at best. Even standard materials may not always be a good criterion for identification of compounds by their luminescence because the color of emission is sensitive to impurity content.

5.4 ENERGY DISPERSION

Energy dispersion is the process of distinguishing the characteristic x-rays from the specimen elements according to their energies rather than their wavelengths. It is based on the proportionality between photon energy and pulse amplitude as discussed in Section 5.1 and depends for its success on the resolution of the detector. Resolution in solid-state detectors (full width at half maximum of the pulse amplitude distribution) is limited to about 200 eV by electronic noise. This would match the resolution of gas proportional counters down to about oxygen K_α but the cryogenic cooling required for solid-state detectors usually means that a window is needed in front of the detector and this limits their use to elements of atomic number about that of magnesium or greater.

In energy dispersion the detector views the specimen directly and measures all of the characteristic x-rays simultaneously. This is especially valuable in electron probe analysis because one can move rapidly

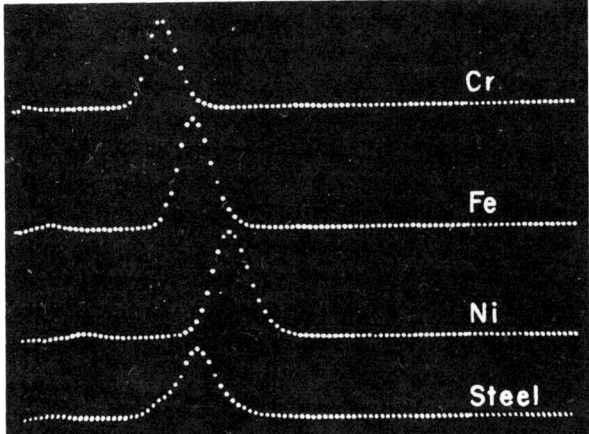

Fig. 5.9 Pulse amplitude distributions from chromium, iron, nickel, and steel as they appear on a multichannel analyzer (proportional counter).

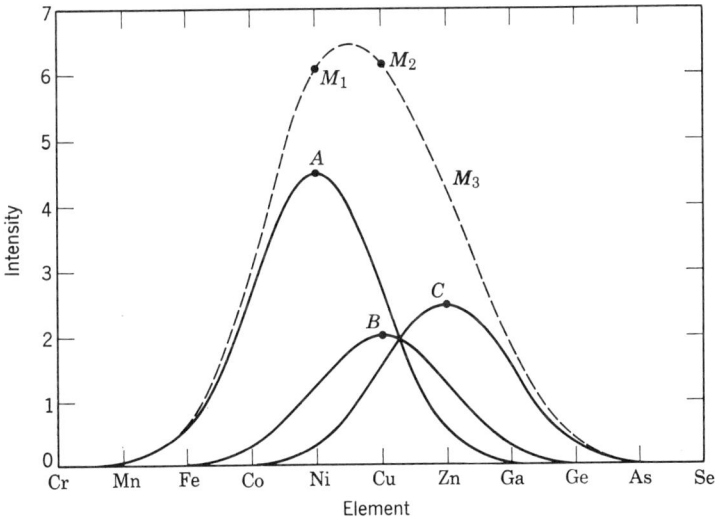

Fig. 5.10 Idealized pulse-amplitude distributions A, B, and C of arbitrary intensities from nickel, copper, and zinc. The dotted curve is the summation of curves A, B, and C.

from place to place on an unknown specimen without worrying about scanning the spectrometer to see what elements are present. The energy spectra are displayed on a multichannel analyzer and appear as shown in Figure 5.9 (taken with a gas proportional counter).

For quantitative analysis it may be necessary to unfold the over-

68   DETECTORS AND ENERGY DISPERSION

lapping energy spectra. For instance, Figure 5.9 shows individual pulse amplitude distributions from chromium, iron, and nickel, and the unresolved energy spectra from a steel sample containing the three elements. The simplest unfolding technique (38, 39), is illustrated in Figure 5.10. Here three component pulse amplitude distributions are represented by A, B, and C and the unresolved composite, M, by the dashed curve. Mathematically, the intensity of the composite at each point is given by the sum of the individual curves. The minimum number of points which must be measured on the composite is equal to the number of individual contributions, in this case three. The best points to choose correspond to the peaks of the individual curves and are shown as M_1, M_2, M_3. For unfolding it is first necessary to know the intensities of the distributions from the pure elements A, B, and C at each of the three positions; these are designated at I_{AA}, I_{AB}, I_{AC}, I_{BA}, I_{BB}, and so

Table 5.2. Analysis of Two Precipitates Such as Those Shown in Figure 5.11

Sample	Measured intensity at the position indicated		
	Iron	Copper	Zinc
Fe	10765	183	32
Cu	749	9730	5439
Zn	541	3970	7845
ppt #1[a]	1015	659	680
ppt #2	1610	740	630

[a] Equations for ppt #1

$1015 = 10765\ R_{Fe} + 749\ R_{Cu} + 541\ R_{Zn}$
$659 = 183\ R_{Fe} + 9730\ R_{Cu} + 3970\ R_{Zn}$
$680 = 32\ R_{Fe} + 5439\ R_{Cu} + 7845\ R_{Zn}$

Table 5.3. Comparison of R Values

Sample	Unfolding			Crystal spectrometer		
	Fe	Cu	Zn	Fe	Cu	Zn
ppt #1	8.8	4.3	5.6	8.6	4.2	4.9
ppt #2	14.3	5.7	4.0	14.2	5.8	4.0

on. One may then write simultaneous equations and solve for relative x-ray intensities of the elements in the composite.

$$\left.\begin{array}{l}I_{MA} = R_A I_{AA} + R_B I_{BA} + R_C I_{CA}\\ I_{MB} = R_A I_{AB} + R_B I_{BB} + R_C I_{CB}\\ I_{MC} = R_A I_{AC} + R_B I_{BC} + R_C I_{CC}\end{array}\right\} \quad (5.3)$$

Tables 5.2 and 5.3 show how the method is applied for the standards and precipitates such as shown in Figure 5.11 and compares the calculated R values with R values measured with a crystal spectrometer on the same precipitates.

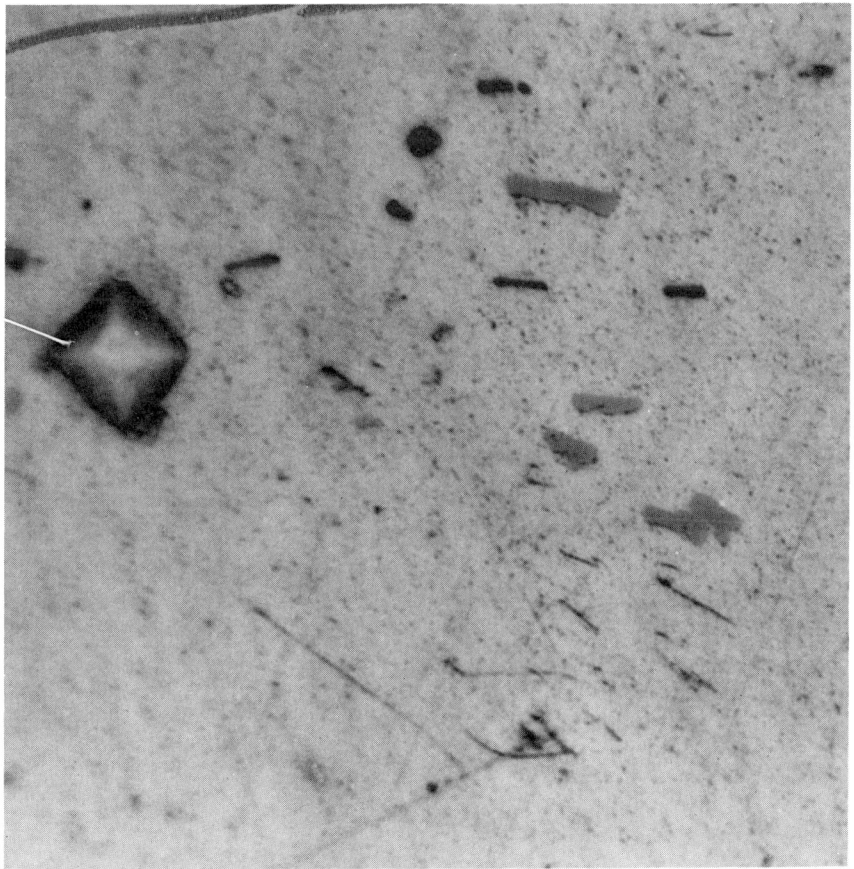

Fig. 5.11 Precipitates in an aluminum alloy. Note the square microindenter mark used to locate the particular region of interest.

It should be noted that energy dispersion and unfolding is not well suited to quantitative analysis of low concentrations, say below 3 or 4%, because the shift in pulse amplitude with counting rate (Section 5.1) depends on the total counting rate rather than that of the individual element. Thus there may be an appreciable (but unknown) shift in the pulse amplitude distribution of a low concentration element compared to the standard for that element; this makes unfolding of the low concentration element subject to relatively large errors.

The simplest form of energy dispersion may be used with crystal spectrometers to eliminate higher order diffraction (Section 4.1). Here the base line and window are set around the desired energy, E (wavelength) and the second- and third-order diffraction of energies, $2E$ and $3E$ ($\lambda/2$ and $\lambda/3$), are eliminated.

CHAPTER

6

TYPES OF SPECIMENS, PREPARATION, EXAMINATION, AND INTERPRETATION

6.1 CATEGORIES OF SPECIMENS

Specimens may be categorized in a variety of ways as shown in Table 6.1. Preparation of specimens generally depends on the nature of the material, while the type of information desired depends on the type of specimen, and the interpretation of results depends on the type of analysis. As far as the electron probe analysis is concerned the field of interest has little bearing except as it affects the other three categories. The sections in this chapter are not chosen from any single category.

6.2 FACTORS IN SPECIMEN PREPARATION

The electron probe microanalyzer measures the characteristic x-rays from a layer a few microns thick on the specimen surface. If the radiation is to be meaningful this layer must be representative of the volume, whether the volume is a precipitate or the bulk matrix. Several factors about the surface are more critical in electron probe analysis than in metallography or x-ray fluorescence. These factors include flatness of surface, etching, electrical conductivity, x-ray take-off angle and secondary fluorescence. One question that is often asked about the specimen is whether there is much temperature rise at the point where the electrons strike. Castaing (40) has found that for a copper specimen with 30 keV electrons and a beam current of 0.5×10^{-6} A, the temperature rise is only 18°C. For nonconductors such as plastics, on the other hand, the energy is not easily dissipated and the temperature rise may be greater than 100°C and may melt the plastic.

6.3 HARD MATERIALS

Hard materials such as alloys, minerals, bone, glass, and ceramics are prepared by sawing and grinding on abrasive papers and wheels or laps. Diamond abrasive is usually the best material for final stages of polishing because the diamond particles will cut hard portions of

Table 6.1. Categories of Specimens[a]

By field of interest	By type of specimen	By type of analysis	By nature of material
Alloys	Inclusions	Identification	Hard
Minerals	Precipitates	Qualitative	Soft
Glass	Phases	Semiquantitative	Particulate
Ceramics	Enriched zones	Quantitative	
Biologicals	Thin films		
Rubber	Low concentration		
Plastics	Low Z elements		
Wood			
Solid-state circuits			

[a] There is no relationship between columns.

the specimen nearly as fast as soft portions and thus produce a flatter surface than the usual Al_2O_3 on felt. Wheel polishing should be minimized and should only be used to remove minor scratches left by the grinding. Overpolishing on wheels tends to smear the surface layer and may result in completely erroneous chemical analysis. Repeated etching and repolishing, as is often done in metallurgical practice, should be avoided because etchants, by their very nature, selectively remove some components and leave a nonrepresentative surface for electron probe analysis. Etching is particularly harmful at grain boundaries, where it may not only remove some constituents but may also redeposit other elements. For example, Marble's etch contains $CuSO_4$, and test specimens etched with it always show strong a copper concentration at grain boundaries. Nonconducting specimens must be coated with a few hundred angstroms of evaporated metal (Al, Cu, Cr, Au) or carbon to prevent charging in the electron beam. This layer reduces the intensity slightly so it is necessary to coat the standards also, even if they are conducting, in order to achieve proper quantitative analysis. Even with a conducting coating, strong insulators may not give the proper relative intensity because of internal charging; a sample of zircon ($ZrSiO_4$) for instance, could not be made to yield more than 92% of the proper zircon intensity when aluminum coated with a range of thicknesses.

Poor electrical conductors are also usually poor thermal conductors and it may be necessary to work at low beam currents to avoid overheating the specimen. Adler (41) found, for instance, a steady decrease in x-ray intensity as sulfur was evaporated from the surface of a fluorite (CaF_2) sample by local heating. Other instances of thermal diffusion toward or away from the irradiated area have also been observed (42).

Gross regions of interest are usually known from preliminary metallography or microscopy. Final selection of the exact area for analysis is made by one of the optical, electron, or x-ray viewing methods discussed in Chapters 1 and 5. If the nature of the specimen is unknown, preliminary etching and examination by a microscope will delineate probable areas of interest. Marking the surface with a diamond indenter and taking micrographs of the marked surface are very helpful in finding the desired areas after the etch has been removed and the specimen has been placed in the electron probe. Figure 6.1 shows a grain boundary diffusion region in Ti–Nb and the square indenter marks used to locate the region in the electron probe (see also Figs. 3.1 and 5.11). The marking

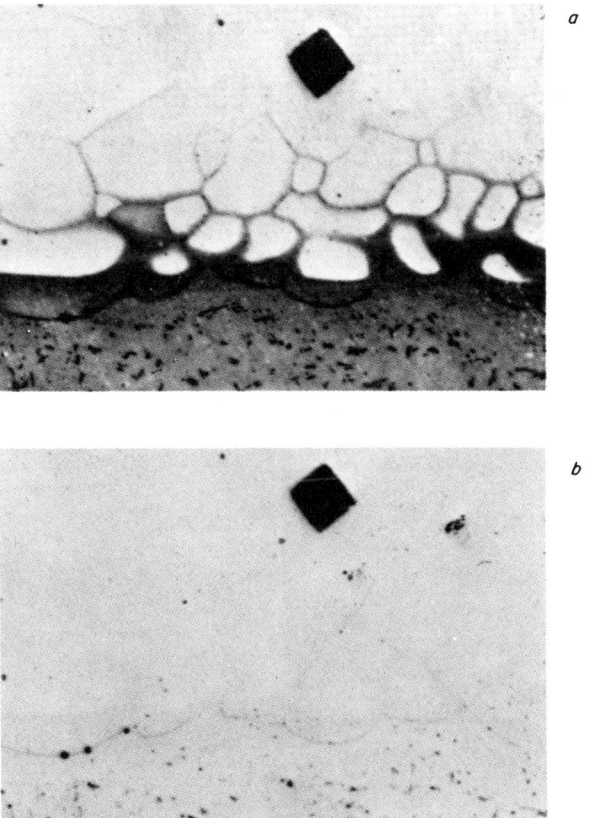

Fig. 6.1 (a) Anodized and punch-marked area of grain boundary diffusion in niobium–titanium diffusion zone. (b) Same area after anodized layer has been removed by light repolishing.

74 TYPES OF SPECIMENS, PREPARATION, ETC.

technique is also of particular value in minerals, where features of interest are often only observable with polarized light.

Qualitative analysis of hard specimens is usually by scanning (Fig. 6.2), but precautions must be considered because of continuum x-ray intensity discussed in Section 1.3. Quantitative analysis is often done using comparison standards such as known phases because it is difficult to obtain pure element standards such as oxygen, sulfur, and so on. Nevertheless the mathematical methods described in Chapters 7 and

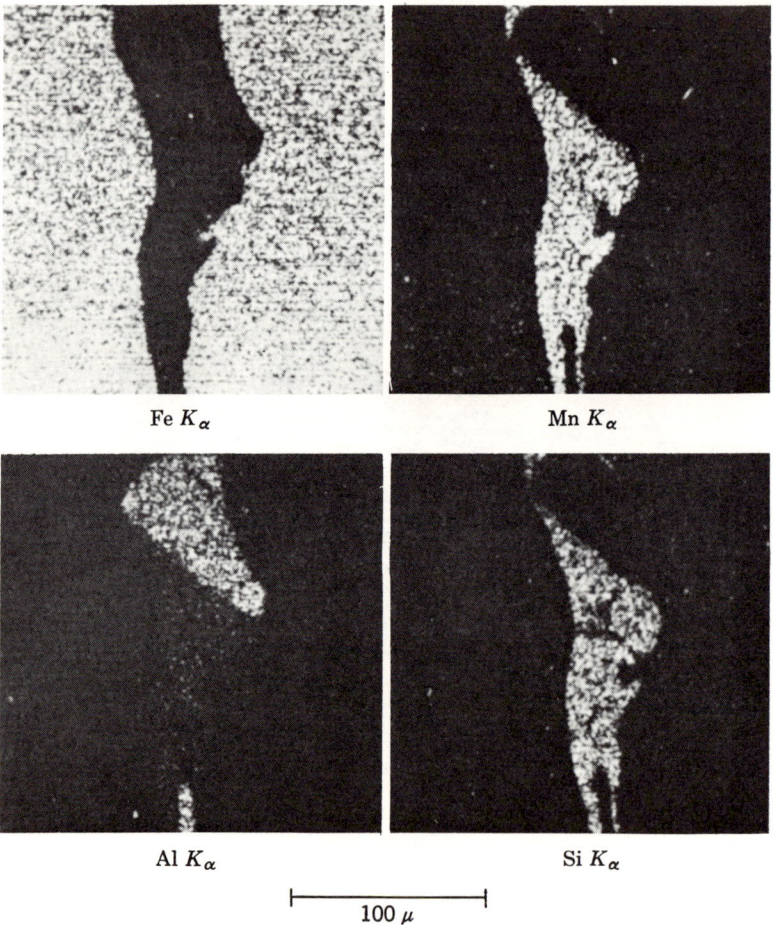

Fig. 6.2 Display pictures taken with the mechanical scanning arrangement of the CAMECA electron probe. The specimen is 1% chromium steel and the vitreous inclusion is rich in manganese with some silicon and aluminum. (Reprinted by permission of CAMECA.)

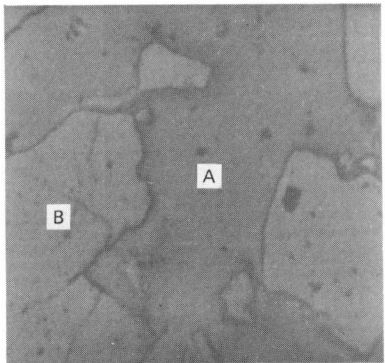
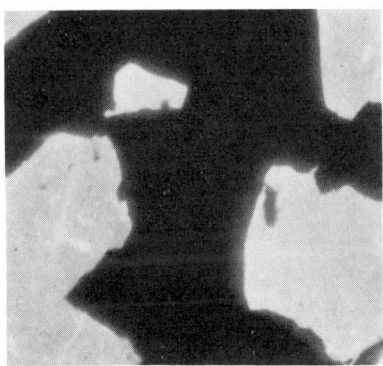

Fig. 6.3 Magnetic grains, A, in sheetlite matrix, B. The top figure is by reflected light and the bottom figure is by electron luminescence. Courtesy Technisch Physische Dienst TNO Delft, Netherlands.

8 are now powerful enough for adequate quantitative analysis of most minerals and similar materials.

Luminescence is useful for distinguishing different minerals. Figure 6.3 shows magnetite grain, A, in a sheelite matrix, B. By reflected light there is little contrast but electroluminescence makes the sheelite ($CaWO_4$) bright blue-white leaving the magnetite (Fe_3O_4) dark.

6.4 SOFT MATERIALS

Soft materials such as biological tissues, vegetable matter, rubber, plastic, or wood are prepared by sectioning with a microtome. Freeze-drying and embedding with resin (43) may be necessary in order to retain the original configuration. Staining with some element which is

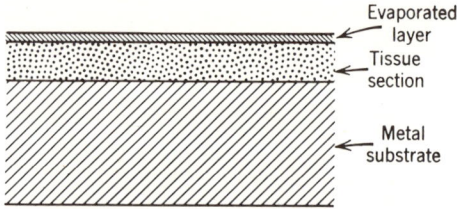

Fig. 6.4 Mounting of tissue section.

readily measured (such as osmium) may be of help in delineating regions of interest. Most of the soft materials are poor electrical conductors and should be placed on a conducting substrate such as metal,* a carbon block, or carbon-coated quartz as well as surface coated with carbon or metal (a transparent substrate has the advantage that light may be transmitted through it for conventional viewing). Figure 6.4 shows such a specimen mounting schematically. Figure 6.5 shows scanning pictures of a tissue section prepared in the above manner. Poor electrical conductors are usually poor thermal conductors as has been mentioned and it may be necessary to work at low beam currents to avoid damaging the specimen.

Often the elements of interest in soft materials are metallic elements which are concentrated in cell walls, membranes, or precipitates. Qualitative analysis by scanning display is very helpful, but because of the low atomic number matrix one must be careful not to misinterpret the results. For instance, if one is looking for copper and has the spectrometer for the scanning display set at the CuK_α line, any other middle or high Z element will give an increase in response in the copper channel because its continuum radiation is more intense than that of the low Z matrix. This is merely an increase in background intensity but can easily be misinterpreted as the presence of copper. The only safe procedure is to stop the beam at such areas and scan the spectrometer off the copper line to make sure it is copper radiation rather than increased background intensity. This point cannot be overemphasized especially with the present tendency to bias out low average intensity and to display the change in intensity from the different spectrometers in terms of superimposed color pictures so that each element is supposedly depicted by a particular color. One can easily imagine situations where

---

* If aluminum is used it should be 99.99% purity or better. This may be chemically polished as follows. Add 5 ml of conc HNO_3 to 95 ml of conc H_3PO_4. Immerse aluminum for 3–15 min with agitation to remove bubbles. Note that 100 ml of solution will polish about 6 to 7 cm^2 of aluminum.

the obvious appearance would be exactly wrong—consider the fictitious case where one may suspect potassium enrichment of cell membranes. Suppose a scanning display in terms of potassium does indeed show an apparent concentration in the cell walls. If the situation is really one of reduced potassium in cell walls but enrichment of some other element, say strontium, the display from the potassium spectrometer would still show increased intensity at the cell walls because of the overwhelming influence of the continuum from strontium. Only by moving the potassium spectrometer off the potassium peak position would one realize that the potassium line was actually weaker at the cell wall. A practical example of such a situation is shown in Figure 6.6 for platinum precipitation in glass. The precipitate shows apparent enrichment in zinc and iron as well as at the background wavelength, 1.86 Å. The intensity scan of Figure 6.6 shows only platinum in the precipitate.

Quantitative analysis for biologicals, vegetable matter, rubber, plastic,

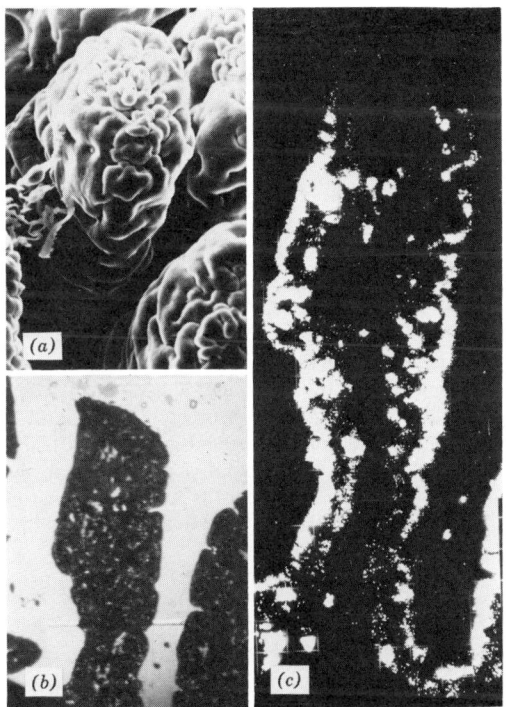

Fig. 6.5 (a) Scanning electron micrograph of a villus in a mammal's small intestine; (b) section of the villus by electron backscatter; (c) same section by iron x-rays. (Courtesy A. J. Tousimis, Biodynamics Research Corp.)

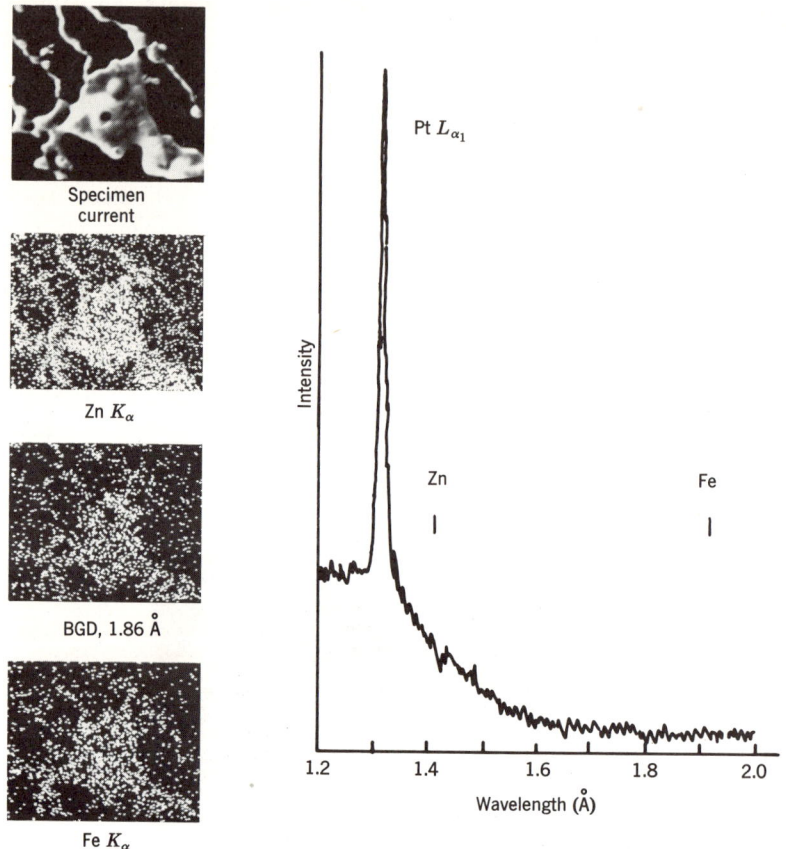

Fig. 6.6 An example of data which can easily be misinterpreted. The platinum precipitate in glass shows clearly in the specimen current display and in the intensity vs. wavelength scan (right). When the spectrometer is set at the iron or zinc position the display pictures could be misinterpreted as iron and zinc enrichment in the precipitate but the display at the background position 1.86 Å shows the same apparent enrichment. The scan shows that there is no iron or zinc present in the precipitate. (Courtesy E. J. Brooks, Naval Research Laboratory.)

and wood is often more difficult than for other solids because the local areas of interest may be smaller than the effective source size and the sections are too thin to stop all of the electrons. This means that the equations of Chapter 7 which are written for the situation where the local composition is homogeneous over the effective source size are not applicable. Equations are not yet available for locally inhomogeneous volumes. One type of comparison standard which has been recommended as effective for biological specimens is to prepare layers of fatty acids with metal atoms attached (43). This is done in much the same fashion used for preparing the pseudocrystals for diffracting long wavelength x-rays as illustrated in Figure 6.7. That is, a particular carbon chain length material such as stearic acid is spread in a monolayer on the surface of water containing a dissolved salt of the metal ion desired. A metal atom attaches to the end of each carbon chain as shown in Figure 6.7. A coated substrate is dipped through this monolayer on the surface and forms a double layer as shown as it is withdrawn. A number of layers may be built up by repeated dippings to approximate the thickness of the specimen sections. The relative amount of metal atoms is controlled by choosing different fatty acids to change the number of carbon atoms per chain and hence the number of carbon atoms separating the layers of metal atoms. For use as probe standards these layers of fatty acids must be coated with evaporated carbon to make them conducting, the beam current should be kept at a low value to avoid burning the material, and the sample should be moved constantly under the beam to avoid damage and ensure good quantitative intensities.

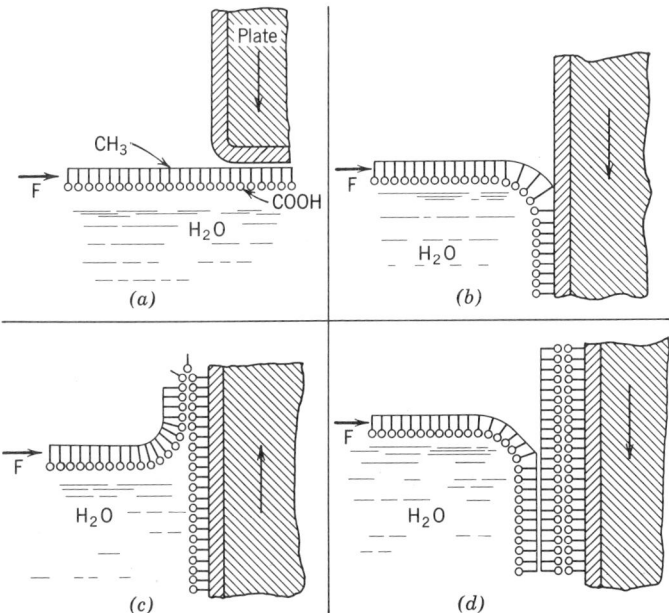

Fig. 6.7 The steps in building up a multilayer pseudocrystal or preparing a comparison standard for biological specimens. (Courtesy A. J. Tousimis, Biodynamics Research Corp., Rockville, Maryland.)

Usually the number of metal atoms is slightly less than the number of chains and the wet chemical analysis as well as the electron probe results are somewhat less than the theoretical composition. According to Tousimis (43) the proper temperature of the water bath varies from one fatty acid to another and must be held to ±1°C for best results.

6.5 INCLUSIONS AND PRECIPITATES

Inclusions and precipitates probably represent the largest single class of specimen examined with the electron probe. They occur in each of the fields of interest listed in Table 6.1. They are often a few microns in size which makes the electron probe about the only suitable tool for examining them individually. They are often more important than the average composition in determining the properties or reactions of materials. The range of interpretation runs from mere identification of major constituents to quantitative analysis of local regions within an inclusion.

Figure 6.8 shows an inclusion in NBS #461 low alloy steel and scanning displays in terms of lanthanum, manganese, and titanium. Scanning dis-

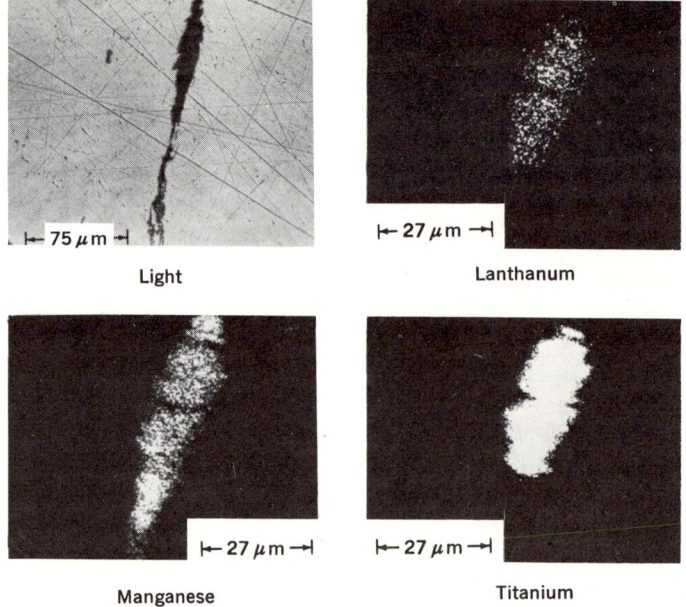

Fig. 6.8 Scanning displays of a precipitate in NBS #461 alloy. (Courtesy K. F. J. Heinrich, National Bureau of Standards.)

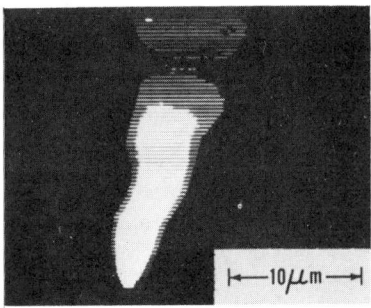

Fig. 6.9 A more quantitative display of Mn in the lower portion of precipitate shown in Fig. 6.8 (Courtesy K. F. J. Heinrich, National Bureau of Standards.)

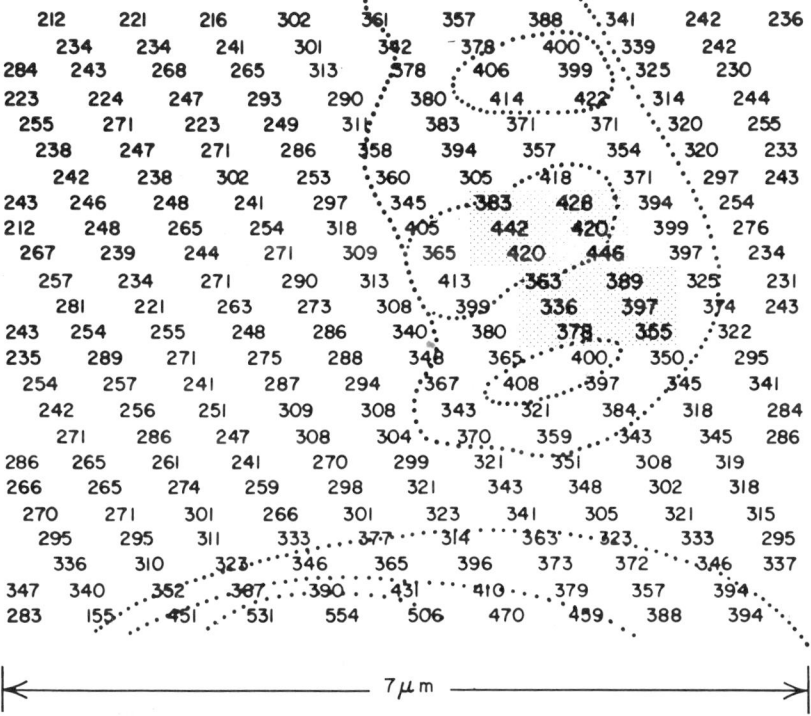

Fig. 6.10 Print-out of intensity topograph to show iron-rich inclusion in aluminum alloy. Numbers are proportional to intensity. Areas outlined manually by connecting equal intensities. Full width of area is 7 μm.

81

plays may be made more quantitative by entering the x-ray intensity into rate-meter circuits whose output voltage changes only at selected rates. Figure 6.9 shows the same inclusion of Figure 6.8 mapped in four levels of manganese content. On the other hand the varying x-ray intensity during scanning may be stored in the memory elements of a multichannel analyzer so that each channel represents the information from one small part of the scan. Printing this out as shown in Figure 6.10 for precipitates in an aluminum alloy gives the most quantitative type of display; areas of various intensity levels may be outlined or shaded as shown.

Figure 6.11 shows inclusions of sinoite in the Jajh deh Kot Lalu meteorite (44). From the electron probe measurements it was possible to identify this previously unreported phase as silicon oxynitride and to determine its stoichiometry as Si_2N_2O. Cathodoluminescence (Section 5.3), is also a valuable means of observing inclusions in minerals.

Precipitates in alloys differ from random inclusions in that they are

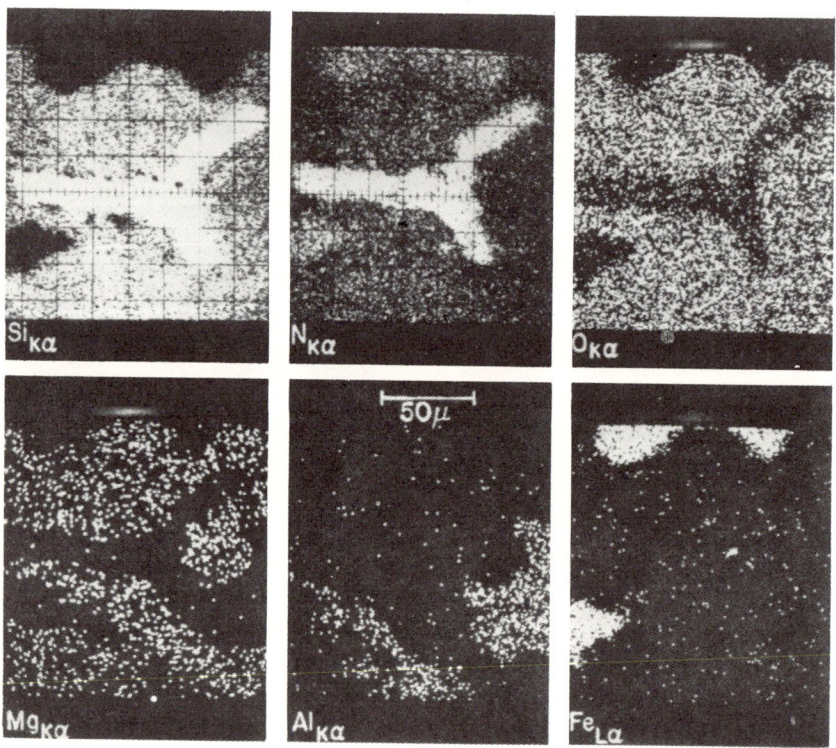

Fig. 6.11

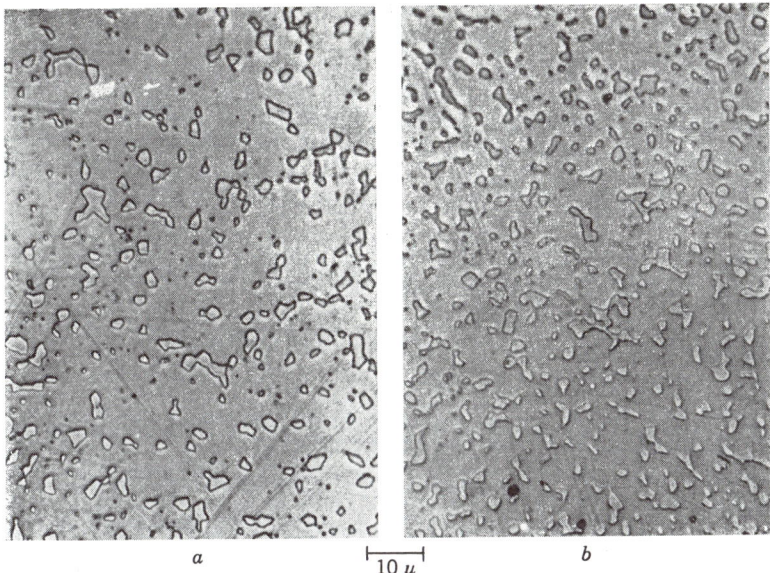

Fig. 6.12 (a) sigma, and (b) chi phases in stainless steel (1).

usually deliberately formed as part of the heat treatment process. Often the likely constituents are known from general reaction theories or experience. What is desired may be the identification of stoichiometric compounds formed in the presence of additional elements in solid solution or the variation of composition with heat treatment Figure 6.12 shows sigma and chi phases in stainless steel; they are indistinguishable in appearance but easily distinguished by the Cr/Mo ratio in electron probe measurements (45). Figure 6.13 shows the results of analysis on a range of precipitate sizes of the chi and sigma phases.

Sometimes it is desirable to know the variation in composition from generally similar precipitates. Figure 5.11 showed precipitates in an aircraft aluminum alloy (46) and Figure 5.9 showed the multichannel analyzer display of the energy spectra from two of the precipitates picked at random. The peaks represent iron, copper, and zinc (the copper and zinc being unresolved with the proportional detector). Table 5.2 showed the variation in iron, copper, and zinc intensities from two of the precipitates. Use of the multichannel analyzer allows very rapid gathering of data from a large number of precipitates so that statistical estimates may be made of compositional variation.

A special technique for extraction of precipitates from an interfering matrix consists of making an extraction replica of the sample as is done

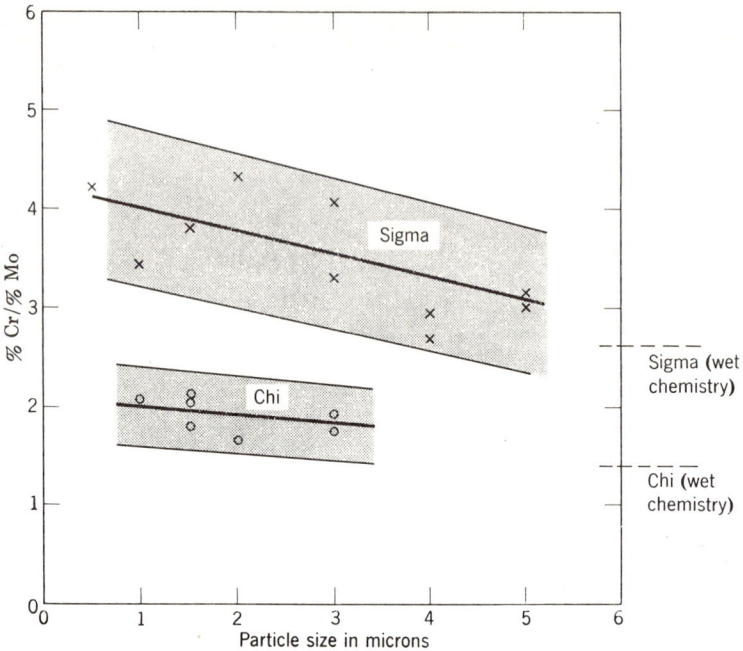

Fig. 6.13 The Cr/Mo ratio allows the sigma and chi phases to be distinguished. The shaded areas are the 95% confidence limits. The electron probe ratios approach the wet chemistry values for extracted residues from specially prepared specimens (1).

for electron microscopy (47). This is especially useful for submicron precipitates containing some of the same elements which occur in the matrix. Figure 6.14 shows the steps in the extraction process schematically. First the specimen is polished in the usual fashion (Fig. 6.14a). Next it is etched and washed to free the precipitates from the matrix being careful not to disturb their position (Fig. 6.14b). Then a block of soft aluminum is pressed against the surface to embed the freed precipitates in the aluminum (Fig. 6.14c). This block of aluminum is the electron probe specimen and looks very much like the original specimen. Figure 6.15 shows an extraction replica from a nickel-base alloy etched with aqua regia; the precipitates are mostly chromium carbides along the grain boundaries. It was not possible to analyze them in the original alloy because of the masking effect of the chromium in the matrix. One precaution should be noted: the particles in the replica will generally not have flat surfaces nor be large enough to give bulk intensities; therefore one cannot obtain accurate intensity measurements relative to bulk specimens of the same material.

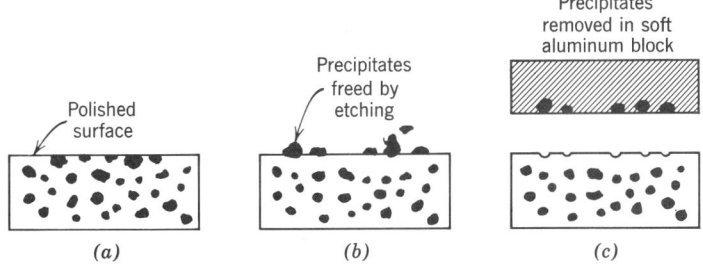

Fig. 6.14 Extraction replica method of sample preparations. (a) Alloy containing precipitates is polished. (b) Surface is etched to free precipitates from matrix without changing their position. (c) Soft aluminum is pressed against etched surface to embed precipitates in the aluminum for electron probe examination.

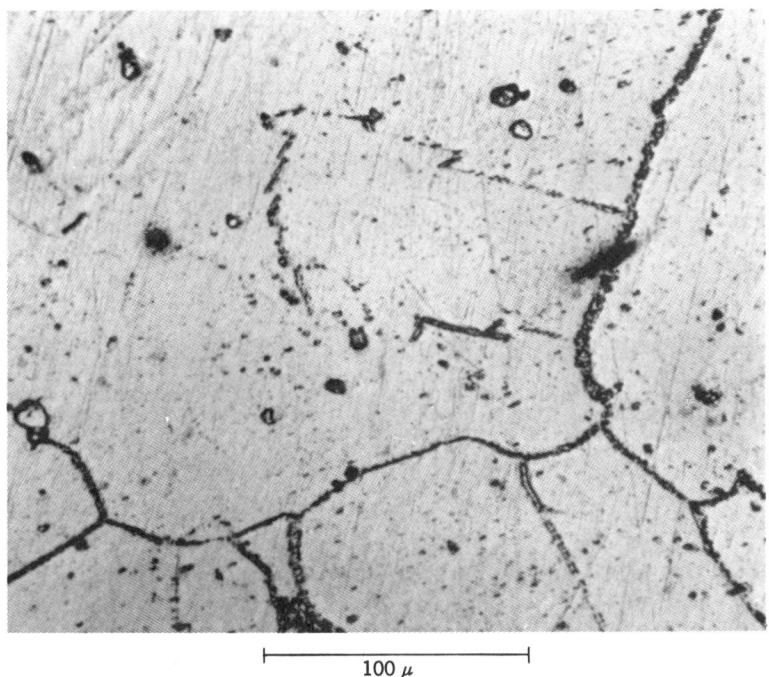

Fig. 6.15 Extraction replica of nickel-base alloy with precipitates along grain boundaries.

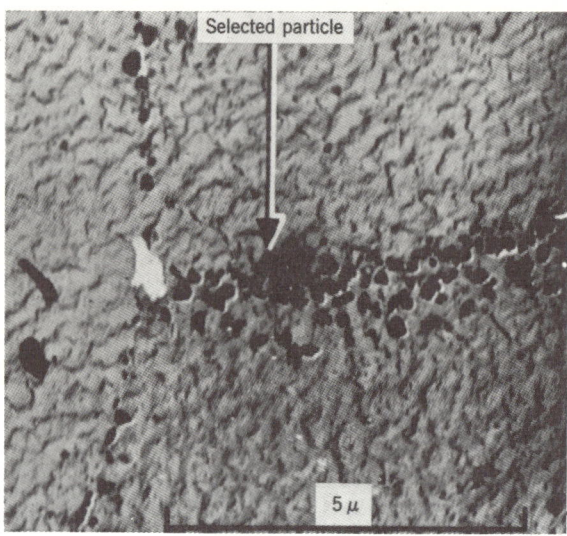

Fig. 6.16 Carbide particles in stainless steel. (Courtesy P. Duncumb, Tube Investments Ltd.)

Some workers prefer to evaporate a carbon film onto the freed precipitates rather than embed them in a block of aluminum. The stripped-off carbon film may then be supported on a grid for electron probe examination. The carbon film method does have the advantage of allowing the precipitates to be observed directly in combination electron probe–electron microscope instruments (see Section 3.5). Duncumb (48) was able to observe 0.1-μm-diameter carbide precipitates from stainless steel by an extraction process as shown in Figure 6.16. He was able to estimate the stoichiometry as $M_{23}C_6$ from a combination of the electron diffraction pattern and the CrK_α and FeK_α measurements on a single 0.3-μm-diameter particle.

6.6 PIGMENTS AND PARTICULATE MATTER

A special technique was devised for preparing samples of pigments from valuable paintings (49). A hypodermic needle was inserted into an area of interest and a core removed. Then the needle was mounted and sectioned as shown in Figure 6.17. It was possible to distinguish the individual pigment particles and to identify them as standard titanium, chromium, or calcium pigments and in at least one case to unmask a forgery because titanium pigments had not been developed at the time the painting had supposedly been made. Particulate matter

Table 6.2. Fallout Particles

Sample No.	Mean diameter (μ)	β-Activity (mμCi)	Color	Shape	First approximation (wt %)							
					Fe	Ca	Si	Al	K	Mn	Ti	P(?)
1	12.0	46.1	Reddish black	Oval	34.3	7.1	16.5	3.2				
2	17.0	12.7	Black	Spherical	33.9	11.0	6.9					
3	13.6	46.1	Reddish black	Spherical	32.3	6.4	5.9	1.4		trace		
4	19.2	4.8	One half: black; another half: white; several red and colorless small particles adhered	Spherical	31.3	10.6	16.8	trace				
5	13.1	6.8	Black and golden, mottled	Semispherical	29.1	5.2	3.2	1.5		0.2	0.7	
6	14.6	50.2	Golden, several angular black small pieces adhered	Spherical	28.2	8.4	3.1			0.2	0.6	

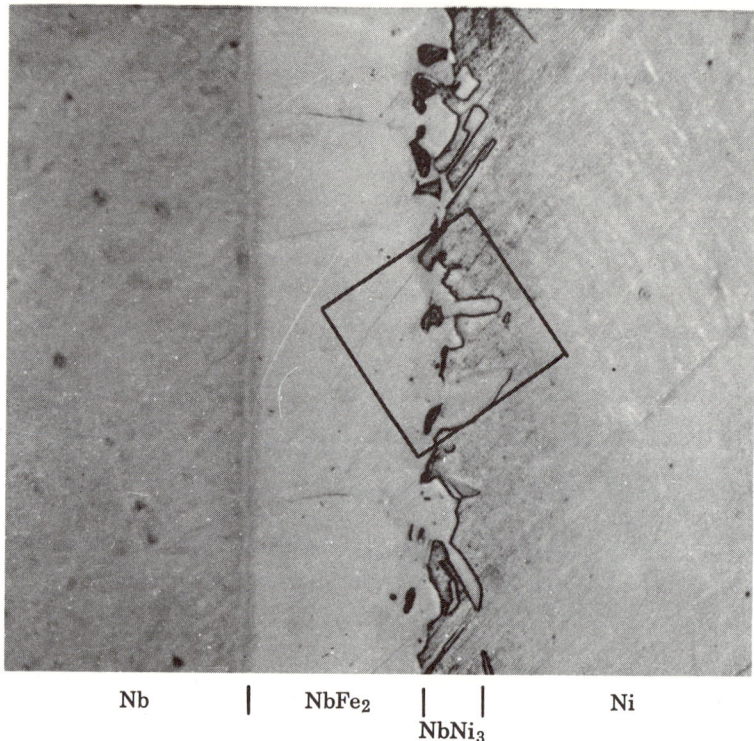

Nb | NbFe₂ | | Ni
NbNi₃

Fig. 6.19 Niobium–iron–nickel ternary diffusion. Niobium and iron were first diffused at 1100°C to form NbFe₂. The excess iron was removed and nickel was bonded to the NbFe₂ zone. Further diffusion at 1100°C resulted in partial replacement of the iron by nickel to form the NbNi₃ needles. The replaced iron then diffused into the parent nickel. The square indicates the region displayed in terms of Fe K_α in Fig. 6.20.

that in the border region and the finger-like extensions, the iron has been almost entirely replaced by nickel combining with niobium.

Often diffusion occurs along grain boundaries as shown in Figure 6.21 for zinc diffusion into copper (51) [Fig. 6.1 showed titanium diffusing into niobium (52)]. Because the grain boundary diffusion rate is as much as 10^6 times faster than the lattice diffusion rate the overall diffusion may be increased by an order of magnitude or more in polycrystalline materials compared to single-crystal material.

Another interesting effect which is readily observed in the electron probe is the influence of impurity reactions. Figure 6.22 shows precipitates of Cr_3P, Cr_2N_5, and CrS which are found when chromium is diffused with iron containing low concentrations of the respective cations (53).

Figure 6.23 shows the effect of low concentrations of oxygen on the solid-solution diffusion in niobium–titanium. In Figure 6.23a, reactor grades of niobium and titanium show the niobium moving a distance of some 600μ into the titanium as indicated by the composition versus distance curves of Figure 6.24a. In Figure 6.23b, a few percent oxygen was first diffused into the niobium and then the niobium was diffused with titanium. The sharp boundary at the line marked A-A may represent stabilization of the αTi phase by the oxygen which has moved quickly out of the niobium into the titanium and then built up in concentration as it is pushed ahead of the incoming niobium. The same feature is shown in Figure 6.24b at 375μ where the niobium content drops rapidly to zero because of the greatly reduced diffusion rate of niobium in αTi.

For the most complete interpretation of diffusion zones, x-ray microdiffraction in combination with electron probe analysis is indicated.

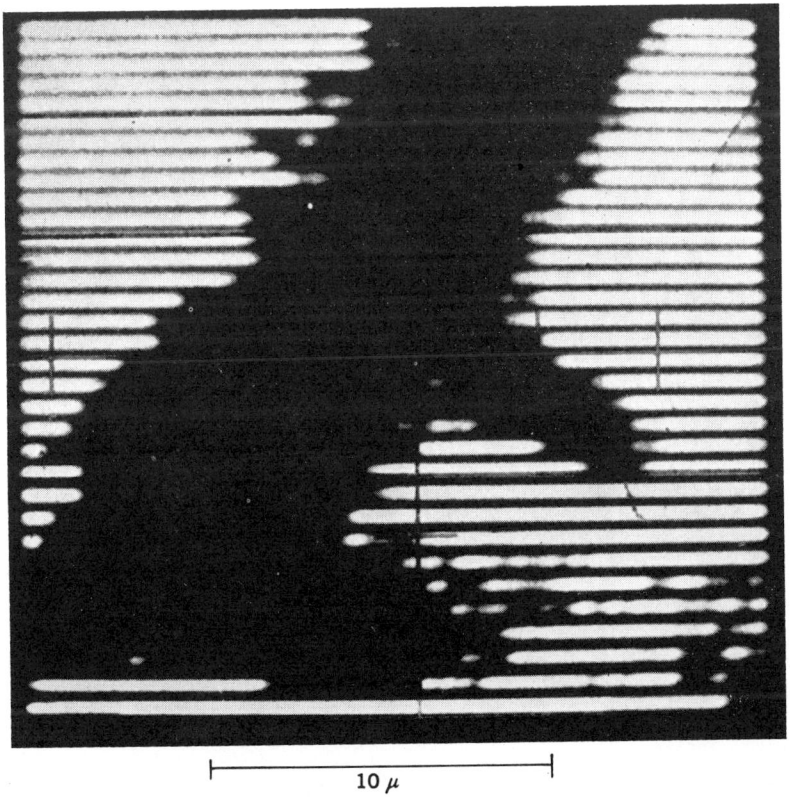

Fig. 6.20 High-magnification display of Fe K_α x-rays in the marked area of Fig. 6.19.

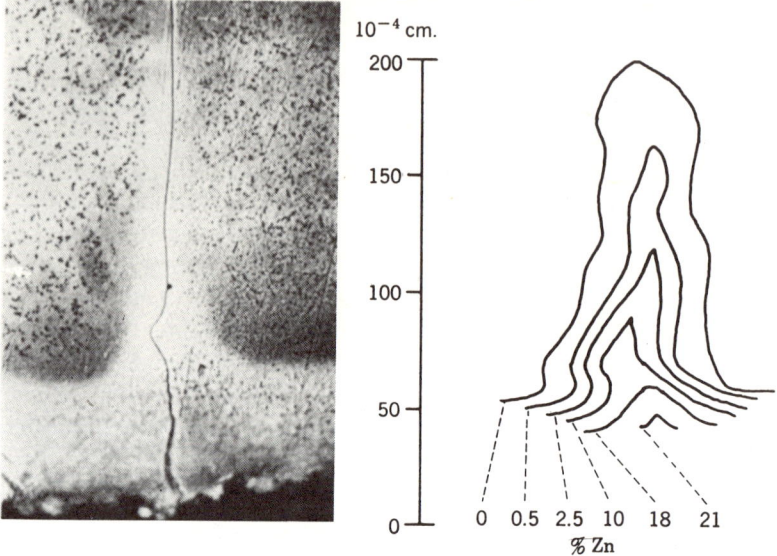

Fig. 6.21 Grain-boundary diffusion of zinc into copper. The contours in the right-hand drawing are isoconcentration lines for zinc (9).

6.8 THIN-FILM SPECIMENS

There are some types of specimens that normally occur as thin films (less than 100 Å thick) and may require special handling. Examples are the electrolytically-thinned metal sections used in electron microscopy, vacuum-evaporated or electroplated metal layers, films for computer memory units, biological sections, and surface corrosion films. The evaporated or plated layers may be examined on the bulk substrate

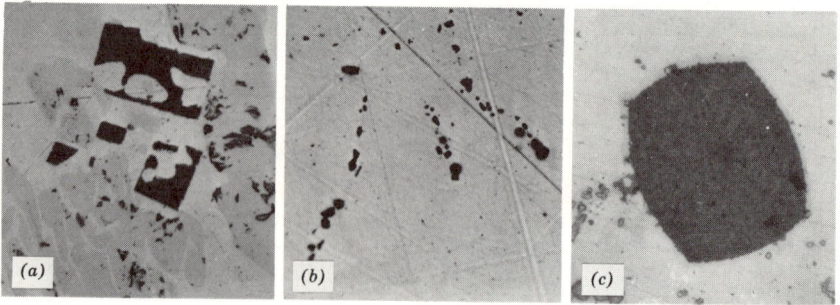

Fig. 6.22 Precipitates in chromium–iron diffusion specimens. (a) Cr_2N_5 (dark squares) and Cr_3P (medium grey patches). (b) CrP along grain boundaries. (c) CrS surrounded by a thin layer of Cr_3P.

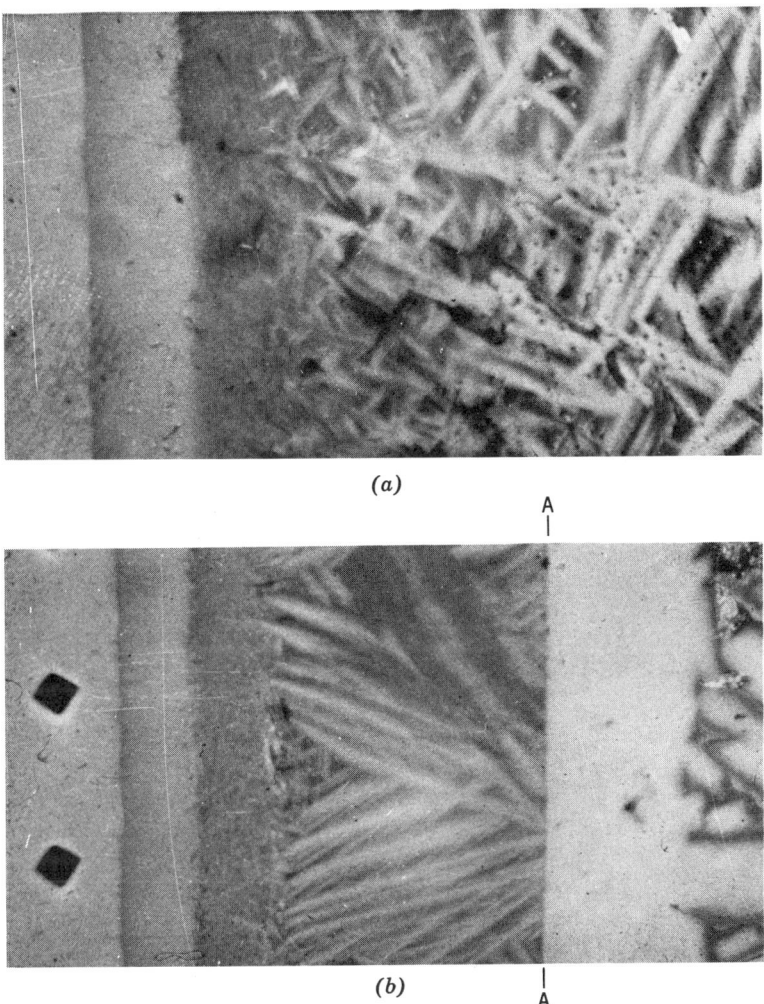

Fig. 6.23 Diffusion zone in (a) Nb–Ti; (b) Nb–(Ti + O₂).

provided it does not contain the same elements as the layers but the thinned sections and stripped-corrosion films will usually require support because of their fragile nature. A wire mesh of the type used in electron microscopy is suitable and may be covered with an evaporated carbon film for extra strength. Pure bulk aluminum (99.99%) may also be used as a substrate.

The incident electron beam will not be completely absorbed in the thin-film specimens (see Fig. 7.4 and Table 7.2 for electron range and source-size information). This results in the emitted x-ray intensity being lower than that from bulk material. Often the reduction in emitted intensity may be used as a measure of the film thickness (55) (Fig. 6.25). Also, electron beams smaller than 1μ in diameter may be used to advantage with thin films because the lateral spread of electrons is approximately equal to the depth of penetration. Thus in a 1000-Å film, the lateral spread will be about 1000 Å, and an initial $0.1\text{-}\mu$ beam will result in a resolution of about 0.2μ. Another advantage of thin-film specimens is that the x-ray intensity is proportional to the mass of the elements present in the film in milligrams per square centimeter. This places quantitative analysis on an absolute basis.

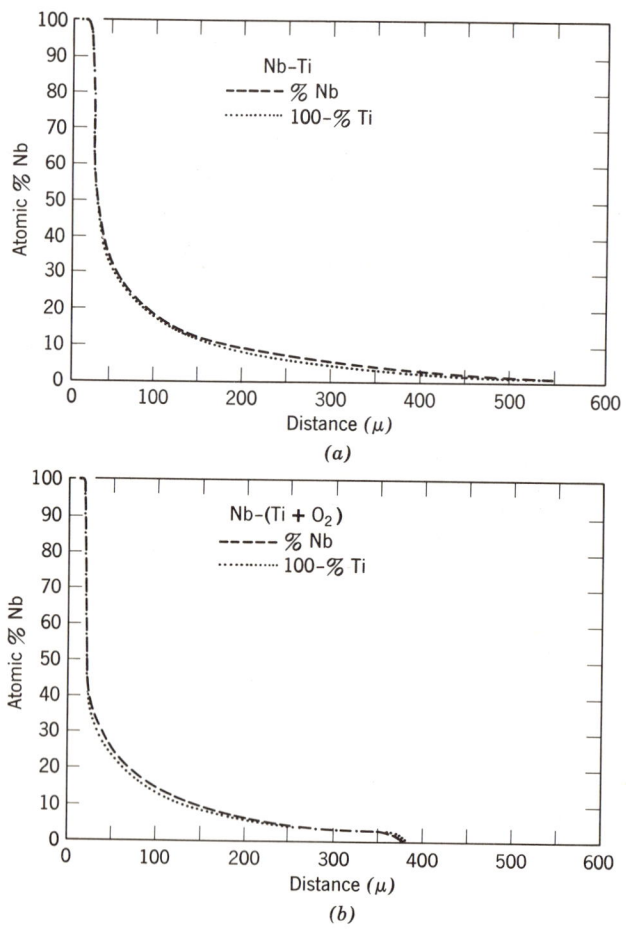

Fig. 6.24 Penetration distance of Nb into Ti.

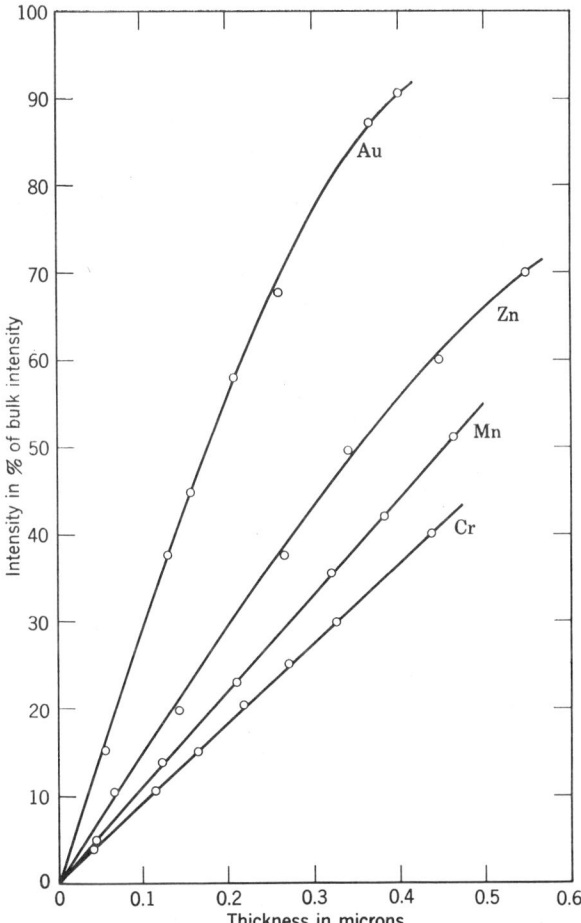

Fig. 6.25 Intensity versus thickness for evaporated metal films.

The limit of detectability for thin films is as small as 10-Å average thickness but, of course, thin layers may not be uniformly thick.

6.9 SOLID-STATE DEVICES

The electron probe is a very powerful tool for studying not only the chemical distribution but also the variation in electrical potential of solid-state devices and circuits. No preparation is required or allowed because these are mostly surface devices. The electron probe is also

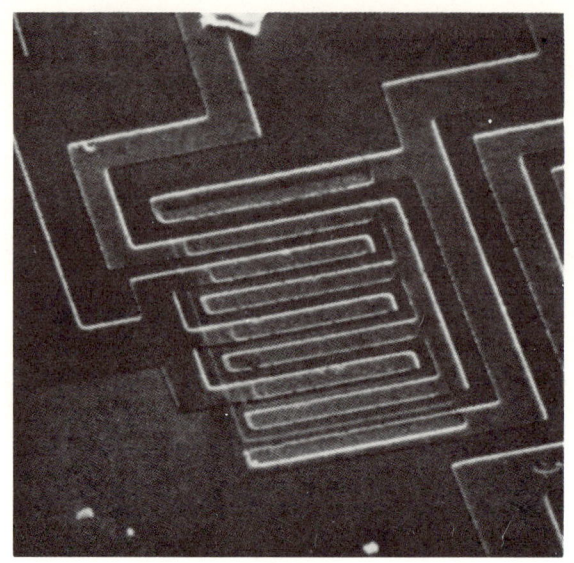

(a)

Fig. 6.26*a*

(b)

Fig. 6.26*b*

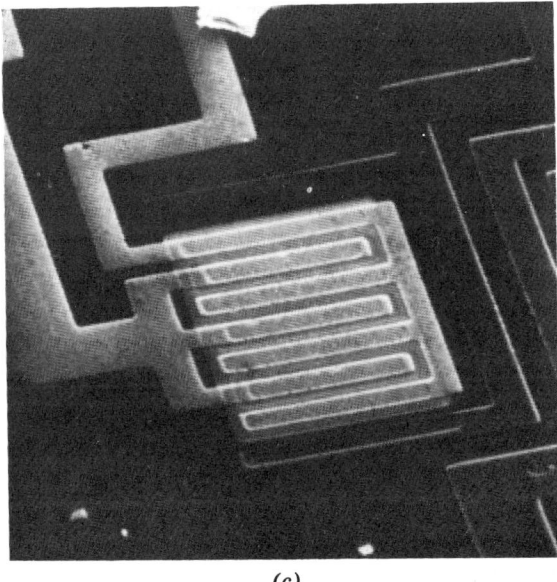

(c)

Fig. 6.26 The effect of voltage applied to solid-state device as observed by secondary back electrons. (a) no bias voltage, (b) emitter bias 5 V, (c) collector bias 20 V. (Courtesy S. Kimoto and J. C. Russ; American Scientist **57**, 124 (1969).

very valuable for identifying the particles of foreign matter which often degrade the device operation and can usually be traced back to faulty manufacturing processes or carelessness in handling which allows dirt to contaminate the surface.

To measure the variation in electrical surface potential the sample is examined by low-energy photoelectrons which are very sensitive to surface potential. Figure 6.26 shows the variation in intensity which results with differing bias on the device during scanning (56). Any flaws in the circuit will show up as unexpected changes in contrast along the graded potential surfaces.

There has been some question about the damage done to the solid-state circuits by electron probe examination. This may be minimized by using reduced beam voltage but its effect should be examined critically in the interpretation of results and use of the circuits after examination. An electron mirror microscope (57) where the incident electron beam does not actually touch the surface may be the best tool, however, for examining surface potential for solid-state circuits but these instruments are not widely available at present.

6.10 ELEMENTS OF ATOMIC NUMBER 11 AND BELOW

Low Z elements pose a special problem for the electron probe because of their low x-ray intensity (because of low fluorescent yield), strong absorption by higher Z elements in the matrix, and the special thin-window detectors required. It is advantageous to reduce beam voltage to 5 kV or even less to reduce the excitation of continuum spectrum and also the excitation of characteristic radiation from higher Z elements; this reduces the general background intensity and makes it easier to measure the weak lines of the low Z elements. In addition, higher

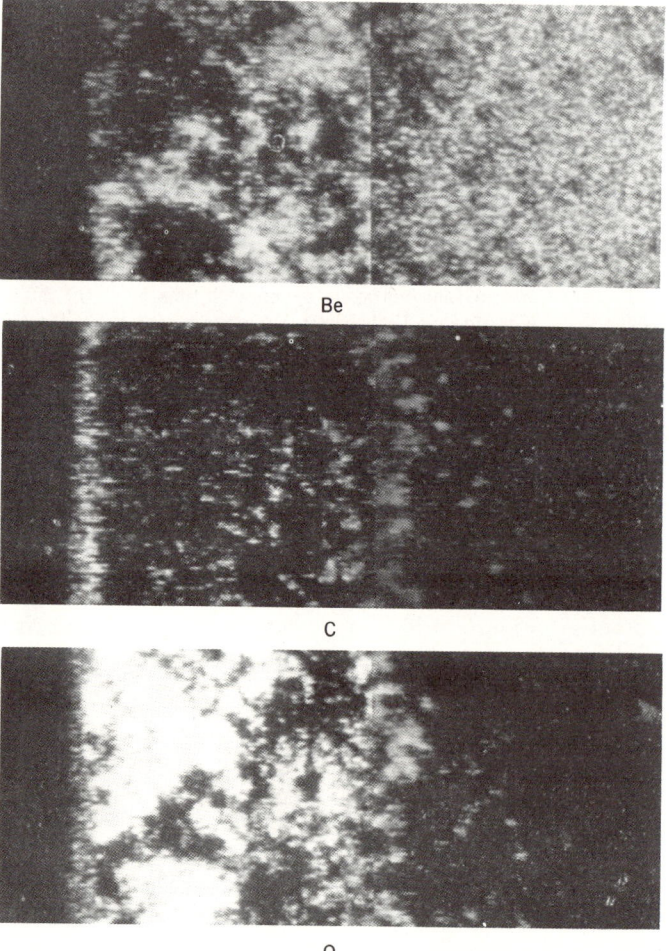

Fig. 6.27 Scanning displays of light elements. (Courtesy R. M. Dolby.)

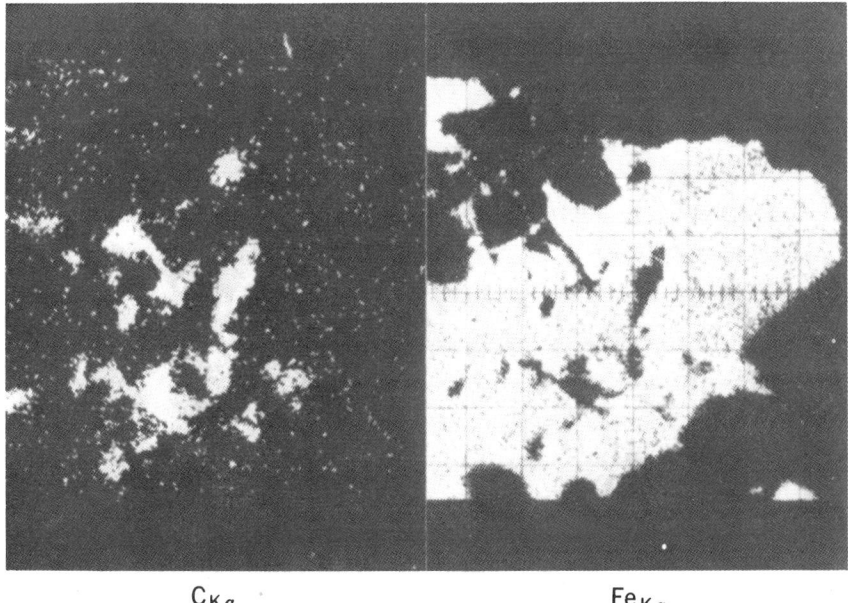

$C_{K\alpha}$ $Fe_{K\alpha}$

Fig. 6.28 Scanning displays of carbon in precipitates. (Courtesy K. Keil and C. A. Anderson.)

operating potential generates the characteristic x-rays from the low Z elements so far below the surface that they are absorbed before emerging while low operating potential generates them near the surface where they can escape. Beam current may be increased to obtain the maximum intensity possible.

With suitable precautions, scanning displays down to beryllium have been obtained (58) as shown in Figure 6.27 and it is commonplace to scan in terms of oxygen or carbon (59) with fairly good resolution (Fig. 6.28).

6.11 MISCELLANEOUS TECHNIQUES IN DISPLAYING AND INTERPRETING SPECIMEN INFORMATION

Very little has been said about viewing through the optical microscope but it is a necessary part of the instrument. The operator uses the microscope to position the specimen in the general area of interest and may be able to recognize interesting inclusions, diffusion zones, and so on. If polarized light is used the optical microscope can delineate birefringent materials. Cathodoluminescence will make some regions easily recognizable in minerals. Thus most probe analysts rely on the optical microscope

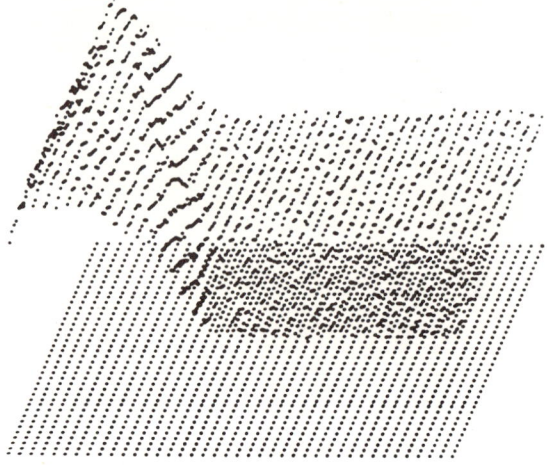

Fig. 6.29 Multiple line-scan display of the edge of a diffusion zone using a multichannel analyzer. (Courtesy K. F. J. Heinrich, National Bureau of Standards, Gaithersburg, Maryland.)

as the first stage in examining the specimen and selecting the particular area to be measured.

Scanning displays in terms of electrons or x-rays have been used throughout this chapter to illustrate the nature of various specimens. A somewhat different type of display is shown in Figure 6.29. Here the x-ray intensity during the scan has been stored in the multiscaler mode of a multichannel analyzer and is displayed with the vertical displacement of each storage channel display representing the number of x-ray photons collected from the corresponding position on the specimen. The appearance is that of a topographic map viewed in perspective and is very effective in depicting the distribution of an element.

Full-color displays are used increasingly for vivid illustrations of the localization of elements. Unfortunately printing costs prohibit color reproductions but the technique is readily imagined. For instance, three scanning displays in terms of three different elements A, B, and C, are photographed. Each of the three negatives is printed through a primary color filter onto an additive material such as color film. Where only A is present, only that color will appear but where A and B are both present the colors will add; for instance, if A is represented by red and B by yellow, A plus B will be orange. The simultaneous display of three elements in a color television tube has been discussed and tried (60) but is not in common use because of expense and color-balancing problems.

CHAPTER

7

INTRODUCTION TO QUANTITATIVE ANALYSIS: EMPIRICAL METHODS; THE CORRECTION-FACTOR APPROACH

Quantitative analysis is the goal of electron probe analysis but is the most controversial subject in the technique. X-ray intensities are not linear with composition because of matrix absorption and enhancement and, to a lesser extent, the variation in absolute x-ray yield with atomic number (the so-called atomic number effects). Referring to Figure 7.1 for the chromium–iron binary system, the iron intensity is reduced below the linear relationship because of strong absorption by chromium whereas the chromium intensity is increased because of secondary fluorescence of chromium by iron radiation. A curve similar to that for chromium in iron would result for chromium in carbon but would be the result of low matrix absorption rather than secondary fluorescence. The matrix effects necessitate the use of comparison standards, calibration curves, or mathematical equations to relate measured intensity to weight percent composition.

In this chapter we will first consider empirical methods and then go on to the correction-factor approach. The following chapter is devoted to more elaborate computer methods such as electron transport which are expected to prevail as high-speed computers become more readily available.

EMPIRICAL METHODS

7.1 CALIBRATION CURVES

The simplest system to analyze is a miscible binary alloy as shown in Figure 7.1. Calibration curves may be prepared by melting known composition standards and measuring the intensities. The method may be extended to limited ternary systems but beyond that it becomes rather costly and time consuming except when the matrix can remain constant and only low concentration additives need be varied. In other solid solution systems it is feasible to chemically analyze a few of the samples

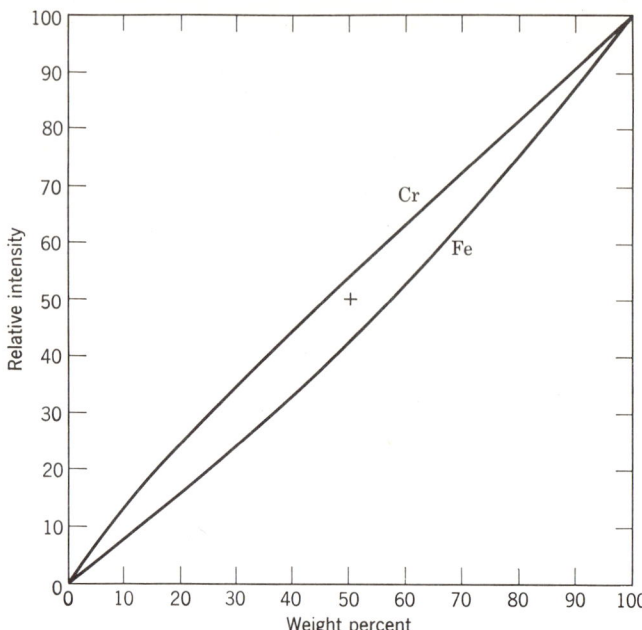

Fig. 7.1 Relative x-ray intensity plotted against weight percent for the chromium-iron system. The FeK_α intensity is reduced below the linear relationship by the strong absorption of chromium. CrK_α intensity is enhanced due to fluorescent excitation by iron radiation.

and use these to prepare calibration curves for one or more constituent. However the method is far less applicable than in x-ray fluorescence analysis where one may mix powders to obtain calibration curves.

7.2 COMPARISON STANDARDS

In the mineral or geological fields, certain phases often recur and it is possible to use these phases in addition to pure elements as standards in analyzing unknowns. For the ideal situation where a standard composition is very similar to the unknown to be measured, one can write

$$C_{iu}/C_{is} = I_{iu}/I_{is}$$

where C_{iu} and C_{is} are the concentration of element i in unknown and standard; I_{iu} and I_{is} are the measured x-ray intensities (above background) of element i in unknown and standard. One of the early electron

Table 7.1. Composition of Copper–Iron Minerals

Mineral	Formula	Iron			Copper		
		X-ray intensity	% X-ray	% Formula	X-ray intensity	% X-ray	% Formula
Chalcopyrite	$CuFeS_2$	126 Hz	std.	30.4	160	std.	34.6
Cubanite	$CuFe_2S_3$	164	40	41.2	98	22	23.4
Valleriite	$Cu_{2-3}Fe_4S_7$	110	26.2	35–39	120	29	20–22
Pyrite	FeS_2	110	std.	46.5	0	—	0
Unknown	$\sim FeS$	212	51[a] 58[b]	63.5?	15–30	0–5[a] 0–5[b]	0

[a] Comparison with chalcopyrite.
[b] Comparison with pyrite.

probe applications (61) to mineralogy concerned the measurement of a phase suspected to be valleriite ($Cu_{2-3}Fe_4S_7$). Intensities of standards and the unknown are shown in Table 7.1 from which it was concluded that the composition was simply FeS and not valleriite at all. A micrograph of that sample is shown in Figure 7.2.

The use of reagent-grade chemical compounds for stanndards is highly questionable because they often contain local inhomogeneities of quite different composition. In fact, most materials except pure elements may be inhomogeneous on a micron-size scale and therefore unsuitable as electron probe standards. Care is required in examining each specific case.

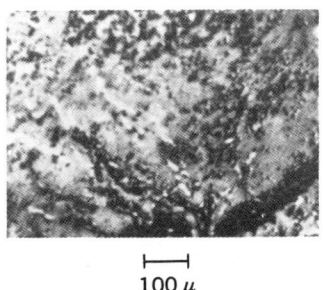

$\longmapsto$
100 μ

Fig. 7.2 The white streaks are an iron rich phase in chalcopyrite ($CuFeS_2$) mineral. Their composition was measured accurately for the first time by using an electron probe They contain about 55% iron and less 5% copper and are probably a sulfide.

7.3 EMPIRICAL COEFFICIENT EQUATIONS

As in the x-ray fluorescence field, it is possible to express interelement effects approximately by empirical coefficients representing the overall effect of one element on the radiation of another element (62). These equations usually take the form

$$k_i = \frac{C_i}{\Sigma \alpha_{ij} C_j} \qquad (7.1)$$

where k_i is the relative x-ray intensity of element i, C is concentration and α_{ij} is the effect on element i by the presence of element j. The α coefficients are obtained by measuring known standards containing the elements of interest; n standards are needed for an n component system, and so on. We will use a three-component system of elements A, B, C to show how the coefficients are determined. Three known standards which we will call 1, 2, 3, containing the elements A, B, C are selected and preferably should contain as wide a range of composition as is expected in analysis of the unknown samples. First we measure k_{A1}, k_{A2}, k_{A3}, that is, the relative x-ray intensity from element A in each of the standards. Then we write the simultaneous equations (these are merely eq. 7.1 with the k and Σ transposed) containing all the known compositions C_{A1}, C_{A2}, C_{A3}, C_{B1}, C_{B2}, and so on.

$$\begin{aligned} C_{A1}/k_{A1} &= \alpha_{AA} C_{A1} + \alpha_{AB} C_{B1} + \alpha_{AC} C_{C1} \\ C_{A2}/k_{A2} &= \alpha_{AA} C_{A2} + \alpha_{AB} C_{B2} + \alpha_{AC} C_{C2} \\ C_{A3}/k_{A3} &= \alpha_{AA} C_{A3} + \alpha_{AB} C_{B3} + \alpha_{AC} C_{C3} \end{aligned} \qquad (7.2)$$

Solving eq. 7.2 for α_{AA}, α_{AB}, α_{AC} may be done by hand for three components but is best done by computer for four or more components.

Next we measure k_{B1}, k_{B2}, k_{B3} in each of the standards and use them in a similar set of equations to obtain α_{BA}, α_{BB}, α_{BC}. The first equation of that set is

$$C_{B1}/k_{B1} = \alpha_{BA} C_{A1} + \alpha_{BB} C_{B1} + \alpha_{BC} C_{C1}$$

and the others may be written by inspection of eq. 7.2. Likewise the coefficients α_{CA}, α_{CB}, α_{CC} are determined by measuring k_{C1}, k_{C2}, k_{C3} and making a third set of simultaneous equations like eq. 7.2.

7.4 USING THE COEFFICIENTS IN ANALYSIS

Once the nine coefficients have been determined from only the three initial standards we may use them to analyze the concentration C_A, C_B, C_C in any unknown sample of the same general type used in determining the coefficients. The procedure is as follows. In the unknown sample, measure k_A, k_B, k_C. Write the set of simultaneous equations containing the nine coefficients, α_{ij} and the unknown composition to be determined, C_A, C_B, C_C.

$$\left.\begin{array}{l}(k_A\alpha_{AA} - 1)C_A + k_A\alpha_{AB}C_B + k_A\alpha_{AC}C_C = 0 \\ k_B\alpha_{BA}C_A + (k_B\alpha_{BB} - 1)C_B + k_C\alpha_{BC}C_C = 0 \\ k_C\alpha_{CA}C_A + k_B\alpha_{CB}C_B + (k_C\alpha_{CC} - 1)C_C = 0\end{array}\right\} \quad (7.3)$$

and

$$C_A + C_B + C_C = \text{constant (see below)} \quad (7.4)$$

These equations are solved for C_A, C_B, C_C. Equation 7.4 is necessary because the simultaneous equations in eq. 7.3 are not mutually independent. The constant in eq. 7.4 is unity if A, B, and C are the only constituents; it may have a value such as 0.98 if the samples have 2% minor constituents which are to be ignored or it may have some other value such as 0.68, 0.83, and so on if A, B, and C are all present as compounds and the compounding element is to be ignored. If the solution of eqs. 7.3 and 7.4 for C_A, C_B, and C_C gives values which do not add up approximately to the constant used in eq. 7.4, a second estimate of the constant in eq. 7.4 may be necessary; a value half-way between the original estimate and the sum of C_A, C_B, and C_C is probably the best second estimate. Then a new solution for C'_A, C'_B, C'_C must be found from eqs. 7.3 and 7.4 and the new values tested again to see if they add up to the estimated constant.

All of the above approach, including the iteration to make the answers consistent with the estimated constant in eq. 7.4, is easily prepared as a computer program and the complete analysis accomplished in one or two minutes. One precaution should be mentioned. The coefficients α_{ij} were determined for a particular electron voltage and must be redetermined if the voltage is changed by more than say 10% because of the variation of the depth at which x-rays are generated (see Section 7.6).

THE CORRECTION FACTOR APPROACH

Castaing, in his thesis (63), outlined mathematical corrections for matrix absorption and fluorescence and gave expressions to be used in converting x-ray intensity to percent composition. Many other workers have introduced variations into the calculations and have modified the expressions for the correction factors. In addition, there are still areas such as geology where analysts often prefer several intermediate composition standards instead of the mathematical calculations which may require only pure element standards or one known standard for each element. It is doubtful that any one author could satisfy all the different view points on the correction factors. The attempt here is rather to present a coherent picture of the factors involved and the generally accepted approaches which are being used.

To introduce the subject, we discuss in general terms electron beam penetration and backscatter, and x-ray generation directly by electrons or by secondary fluorescence. Then we will show the mathematics of the correction-factor approach and the expressions commonly used for the individual correction factors. Computer programs are available from several sources (64) for speeding the calculations. More discussion is found in Section 7.13.

7.5 ELECTRON-BEAM PENETRATION AND BACKSCATTER

Figure 7.3 shows typical electron paths in the specimen schematically. The incident electrons lose energy by three processes of interaction with the atoms.

1. The most common interaction, and a major source of electron energy loss, is scattering by the electron distribution around the atoms; this usually results in only slight changes in direction ($<1°$) and loss of a few eV in energy per scattering event. The mean path length in gold between scatterings is about 0.1 μm for 30 keV electrons and in aluminum is only about 20% shorter.

2. The next most common interaction (but a small contribution to total energy loss compared to scattering) is with the nucleus of the atom. This results in generation of a bremstrahlung x-ray photon. The electron will lose energy corresponding to the photon generated; this may be any value up to the full energy of the electron.

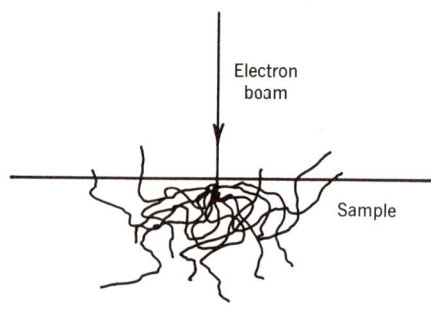

Fig. 7.3 Schematic representation of electron paths in the specimen. Some electrons escape from the surface but others dissipate all their energy in the specimen.

3. Only slightly less likely than bremstrahlung generation is the interaction of removing an inner-shell electron to generate a characteristic x-ray photon. The energy lost in this process goes partly to the x-ray photon and partly to the kinetic energy of the inner-shell electron which is ejected. Electron generation of characteristic x-rays is discussed in detail in Section 7.6.

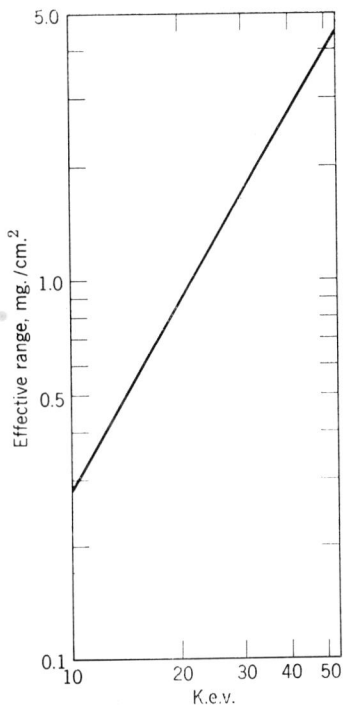

Fig. 7.4 Electron range plotted against energy (8).

Electron range is principally dependent on energy and density and is shown in Figure 7.4. For low Z materials such as biologicals the range will be nearly ten times greater than in steel alloys because of the difference in density. This along with the variation in absorption for emerging x-rays leads to a large variation in effective x-ray source size. Table 7.2 shows source-size data for a variety of conditions.

Electron backscatter fraction depends on atomic number as shown in Figure 7.5. The energy distribution for backscattered electrons is given in Figure 7.6. It is of interest, but unrelated to quantitative analysis,

**Table 7.2. Effective X-Ray Source Size
(95% of X-Ray Generation)**

μ' csc ψ	20 keV	30 keV	40 keV	50 keV
Max. range	0.9 mg/cm²	1.8 mg/cm²	2.95 mg/cm²	4.3 mg/cm²
100	0.9	1.37	1.93	2.80
1000	0.76	1.03	1.50	1.90
5000	0.45	0.54	0.63	0.72

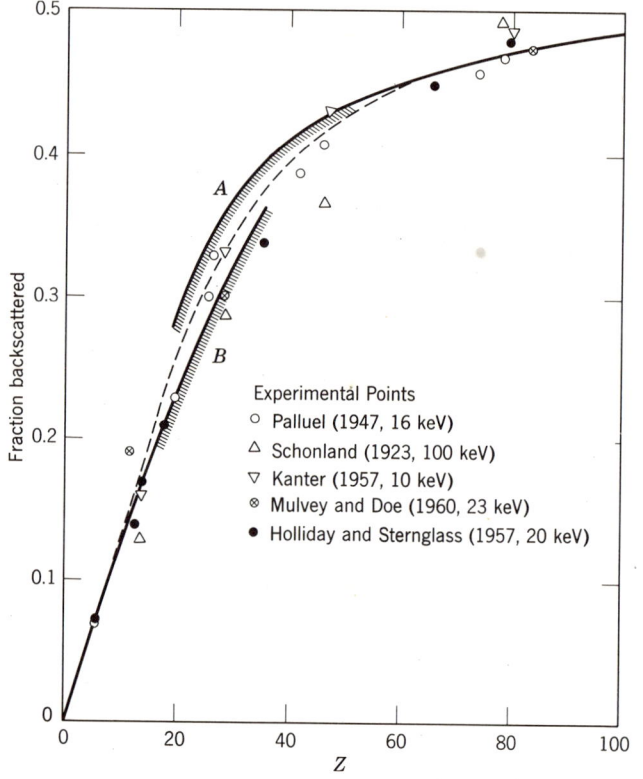

Fig. 7.5 Fraction of electrons backscattered as a function of atomic number.

that if a retarding grid is placed ahead of the backscatter detector and set at about −50 V, it will remove most of the low energy Auger and other secondary electrons. The remaining high-energy electrons, when used for scanning displays, give reasonably good contrast for regions of different average atomic number. On the other hand, it is the low-

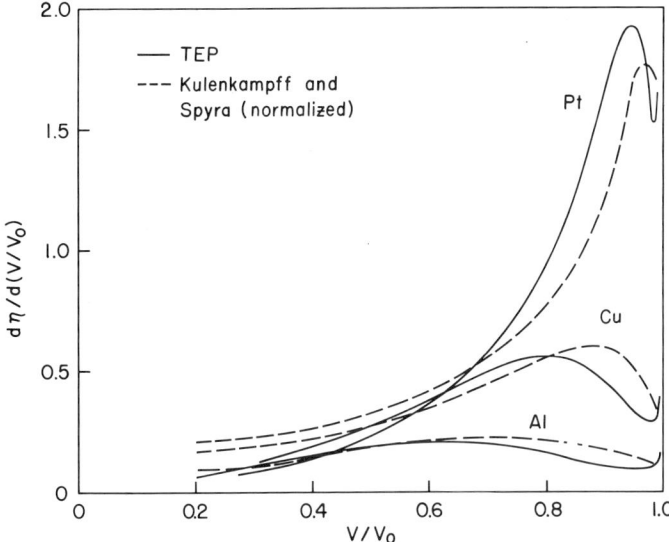

Fig. 7.6 Energy distribution of back scattered electrons, $d\eta/d(V/V_0)$ is a scaled value for the number of electrons of energy V/V_0. [Courtesy *J. Appl. Phys.*, **40**, 1627 (1969).]

energy secondary electrons which are most useful in measuring surface-potential variations in solid-state circuit samples; for these measurements a weak accelerating grid is used on the detector but the detector is placed out of the way of the high-energy backscattering (Fig. 1.1).

7.6 ELECTRON GENERATION OF CHARACTERISTIC X-RAYS

The most important source of characteristic x-rays is direct generation by the electron beam. For quantitative analysis it is necessary to know the x-ray generation as a function of composition, depth below the surface, and incident electron energy. The first measurements of the depth function were made by R. Castaing and J. Descamps *J. Phys. Radium* **16**, 304 (1955) and R. Castaing and J. Henoc *X-ray Optics and Microanalysis* Herman & Cie, Paris, 1966 and are shown in Figure 7.7 along with more recent calculations of D. B. Brown. (78) The measurements were made using the arrangement of Figure 7.8; corrections were made for absorption of the emerging radiation. The term $\phi(\rho z)$ represents the number of x-rays generated in a thin layer, $\Delta(\rho z)$, at depth ρz but normalized by dividing by the number generated in the same layer in free space. The $\phi(\rho z)$ curves are different for different elements and different

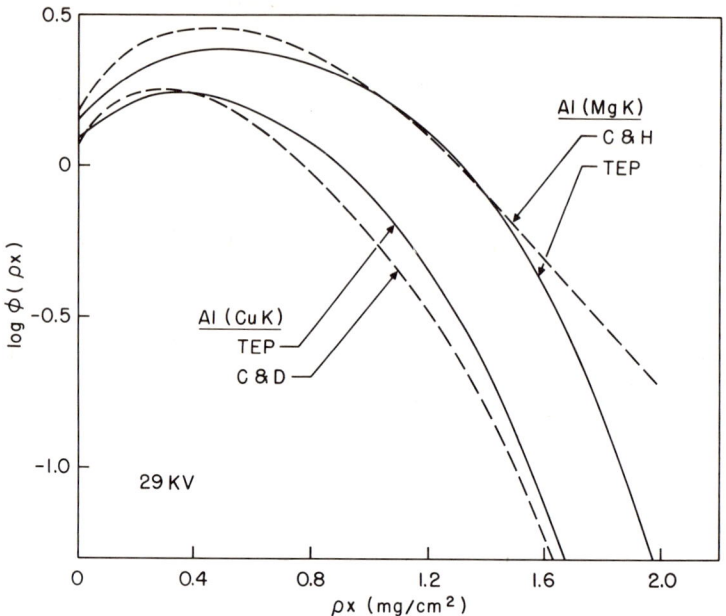

Fig. 7.7 Measured and calculated values of x-ray generation as a function of depth below the surface. [Courtesy *J. Appl. Phys.*, **40**, 1627 (1969).]

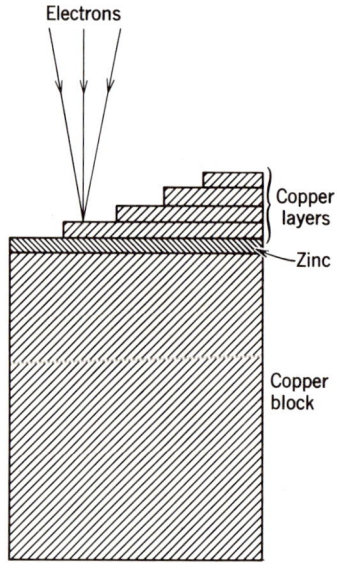

Fig. 7.8 Experimental arrangement for measuring the generation of x-rays as a function of depth below the surface. [Courtesy *J. Phys. Rad.*, **16**, 304 (1955).]

matrix composition. The peak in each curve occurs for the combination of circumstances where (1) the electrons have been deflected enough to increase their path length through layer $\Delta(\rho z)$ and hence the amount of x-ray generation per electron and (2) the energy of the electrons has not yet been reduced much by scattering. Therefore the number of x-ray photons generated per layer increases with depth initially.

If $\phi(\rho z)$ is known it is possible to write the total intensity, I_{Mi}, from element i in matrix M as

$$I_{Mi} \propto C_i \int \phi_{Mi}(\rho z) \exp\left[-\mu_{Mi}\rho z \csc \psi\right] d(\rho z) \qquad (7.5a)$$

where C_i is the concentration (weight fraction) of i; μ_{Mi} is the mass absorption coefficient of matrix M for the characteristic radiation of ment i; and ψ is the take-off angle.

$$\mu_{Mi} = \mu_{1i}C_1 + \mu_{2i}C_2 + \cdots \mu_{ii}C_i + \cdots$$

For a pure element standard the intensity I_{100i} is similarly

$$I_{100i} \propto \int \phi_{100i}(\rho z) \exp\left[-\mu_{ii}\rho z \csc \psi\right] d(\rho z) \qquad (7.5b)$$

where μ_{ii} is the absorption coefficient of element i for its own characteristic radiation (eqs. 7.5a and b have the same proportionality constant).

During the development of quantitative electron probe analysis, it became the practice to consider the expression for intensity as a function of $\mu \csc \psi$ or χ (chi) as it is commonly called, and to define a function $f(\chi)$ as

$$f(\chi) = \frac{\int \phi(\rho z) \exp\left[-\mu\rho z \csc \psi\right] d(\rho z)}{\int \phi(\rho z) d(\rho z)}$$

This corresponds to normalizing, in other words, dividing the emitted intensity by the generated intensity. One may then write the relative x-ray intensity, k_i, as

$$k_i = \frac{I_{Mi}}{I_{100i}} = C_i \left[\frac{f(\chi_{Mi})}{f(\chi_{100i})}\right] \left[\frac{\int \phi_{Mi}(\rho z) d(\rho z)}{\int \phi_{100i}(\rho z) d(\rho z)}\right] \qquad (7.6)$$

The first bracketed term is called the absorption correction term because it contains all the absorption effects. The second bracketed term is called the atomic-number correction term; when multiplied by C_i it represents the absolute intensity generated in element i in matrix M compared (normalized) to the intensity generated in a sample of pure element i. Expressions for these terms are considered in Sections 7.9 and 7.10.

7.7 SECONDARY FLUORESCENCE

CONTINUUM FLUORESCENCE

This term is small and is often neglected in quantitative analysis because of its complexity. The continuous x-ray spectrum generated by the electron beam will in turn excite some characteristic radiation by fluorescence. It is not a simple calculation to estimate this contribution to intensity but Henoc (65, 66) has developed equations which seem to allow an adequate estimate. In addition, the electron transport method (Chapter 8), allows an independent estimate which gives similar answers. In either approach the source of continuum radiation can be assumed, for simplicity, to be entirely at the surface because the electron range is small compared to the penetration of the primary x-rays in exciting secondary fluorescence.

CHARACTERISTIC FLUORESCENCE

The characteristic radiation of one element in the matrix may have enough energy to excite other characteristic radiation of other elements by fluorescence. In fact, this is usually a much greater contribution than continuum fluorescence and is easier to treat mathematically. The primary generation of radiation from element j in layer ρdz is given by the depth distribution $\phi(\rho z)$ for curves like Figure 7.7. The amount of this characteristic radiation which reaches layer $d(\rho y)$ is controlled by the matrix absorption, μ_{Mj}. The fluorescence of element i in layer $d(\rho y)$ is given by the absorption of element i for the radiation from element j and the fluorescent yield for element i. By integrating over all angles ϕ, all layers ρdy and all starting layers ρdz, for each element j in ρdz which will excite element i in layer ρdy, one obtains the total characteristic fluorescence of element i.

7.8 THE CORRECTION FACTOR EQUATION

The correction factor concept, after Duncumb, assumes that relative x-ray intensity, k_i, is equal to weight fraction times a series of correction factors for atomic number, Z, absorption, A, and fluorescence, F.

$$k_i = \frac{I_{Mi}}{I_{100i}} = C_i Z \, A \, F \tag{7.7}$$

Mathematically the terms Z, A, and F cannot be separated so nicely as indicated, and eq. 7.7 is known to be an approximation based on empirical factors and comparison with a limited number of known compositions. Nevertheless it does lend itself to pencil-and-paper evaluation and has also been written in a variety of computer programs to speed the calculations. The separate factors given in the following sections are due to Duncumb and Reed (67) for atomic number, Philibert (68) for absorption [modified by Duncumb and Shields (69)], and Reed (70) for fluorescence.

7.9 THE ABSORPTION CORRECTION FACTOR, A

The absorption factor, A, was expressed in eq. 7.6 as

$$A = \frac{f(\chi_{Mi})}{f(\chi_{100i})}$$

and the value for $f(\chi)$ has been given by Philibert (68) as

$$f(\chi) = \frac{1}{\{(1 + \chi/\sigma)[1 + h\chi/\sigma(1 + h)]\}} \quad (7.8)$$

where the values for χ and h change for the pure element i and the matrix M; $\chi_{100i} = \mu_{ii} \csc \psi$; and $\chi_{Mi} = \mu_{Mi} \csc \psi$.

$$h_i = \frac{1.2 A_i}{Z_i^2}$$

but

$$h_M = \frac{1.2}{\Sigma(C_i Z_i^2/A_i)}$$

where Z is atomic number and A is atomic weight.

The value for σ is that of the element being determined no matter what the matrix. Originally σ was considered as just the Lenard coefficient but its value has been modified by Duncumb and Shields (69) and further by Heinrich (71). The best present estimate is

$$\sigma_i = \frac{4.5 \times 10^5}{(V - V_i)^{1.65}}$$

where V and V_i are the electron-beam voltage and the excitation voltage for element i, respectively (see Section 7.12 for a numerical example of quantitative calculations using the correction factors.

7.10 THE ATOMIC NUMBER CORRECTION FACTOR, Z

In eq. 7.6 the atomic-number term was given as C_i times the ratio of intensity generated in element i in a matrix M divided by the intensity generated in pure element i. To evaluate the ratio of integrals following the approach of Duncumb and Reed (67) we can write

$$Z = \frac{\int \phi_{Mi}(\rho z)\, d(\rho z)}{\int \phi_{100i}(\rho z)\, d(\rho z)} = \frac{R_M \bar{S}_i}{R_i \bar{S}_M} \tag{7.10}$$

where R is called the backscatter coefficient; it is the ratio of ionization actually produced to that which would have been produced if there were no backscatter of electrons. Values of R_i are given in Table 7.3; and

Table 7.3. Values of Backscatter Coefficient R as a Function of 1/U and Z

Z	\multicolumn{11}{c}{1/U}
	0.01	0.10	0.20	0.30	0.40	0.50	0.60	0.70	0.80	0.90	1.00
0.	1.000	1.000	1.000	1.000	1.000	1.000	1.000	1.000	1.000	1.000	1.000
10.	0.934	0.944	0.953	0.961	0.968	0.975	0.981	0.988	0.993	0.997	1.000
20.	0.856	0.873	0.888	0.903	0.917	0.933	0.948	0.963	0.977	0.990	1.000
30.	0.786	0.808	0.828	0.847	0.867	0.888	0.911	0.935	0.959	0.981	1.000
40.	0.735	0.760	0.782	0.804	0.827	0.851	0.878	0.907	0.938	0.970	1.000
50.	0.693	0.718	0.741	0.764	0.789	0.817	0.847	0.881	0.919	0.959	1.000
60.	0.662	0.688	0.713	0.737	0.764	0.793	0.825	0.862	0.904	0.950	1.000
70.	0.635	0.663	0.687	0.713	0.740	0.770	0.805	0.844	0.889	0.941	1.000
80.	0.611	0.639	0.665	0.691	0.718	0.750	0.785	0.826	0.874	0.932	1.000
90.	0.592	0.613	0.639	0.665	0.695	0.730	0.767	0.811	0.862	0.924	1.000
99.	0.578	0.606	0.634	0.661	0.691	0.725	0.763	0.806	0.858	0.921	1.000

$R_M = \Sigma C_i R_i$. The term $\bar{S}$ is the mean electron stopping power and is expressed as

$$\bar{S} = \frac{Z}{A} \ln \frac{1.17 \bar{V}}{J} \tag{7.11}$$

where $\bar{V} = (V + V_i)/2$ in volts and $\bar{S}_M = \Sigma C_i \bar{S}_i$. In eq. 7.11 the value of the term J, the mean ionization potential, has been the subject of some questioning and controversy but the most recent work of Duncumb, Shields-Mason, and da Casa (72) give it in terms of J/Z as

$$\frac{J}{Z} = 14(1 - e^{-0.17}) + \frac{75.5}{Z^{Z/7.5}} - \frac{Z}{(100 + Z)} \tag{7.12}$$

from an empirical fit to the curve of Figure 7.9. Table 7.4 lists the J values for the elements (see Section 7.12 for a numerical example using the correction factors).

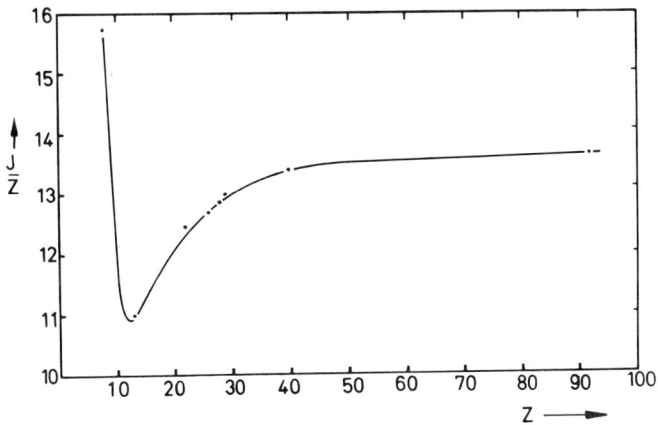

Fig. 7.9 Mean ionization potential J plotted as J/Z against Z determined empirically from a series of microprobe analyses of known alloys.

Table 7.4. Mean Ionization Potential J Determined Empirically as a Function of Atomic Number Z

Z	J(eV)	Z	J(eV)	Z	J(eV)	Z	J(eV)
6	146	29	377	52	706	75	1017
7	135	30	392	53	720	76	1031
8	127	31	407	54	734	77	1044
9	123	32	422	55	747	78	1057
10	123	33	437	56	761	79	1071
11	126	34	451	57	775	80	1084
12	133	35	466	58	788	81	1097
13	142	36	481	59	802	82	1111
14	154	37	495	60	815	83	1124
15	166	38	510	61	829	84	1137
16	180	39	524	62	843	85	1151
17	194	40	538	63	856	86	1164
18	209	41	553	64	870	87	1177
19	224	42	567	65	883	88	1191
20	239	43	581	66	897	89	1204
21	255	44	595	67	910	90	1217
22	270	45	609	68	923	91	1231
23	286	46	623	69	937	92	1244
24	301	47	637	70	950	93	1257
25	316	48	651	71	964	94	1270
26	332	49	665	72	977		
27	347	50	679	73	991		
28	362	51	692	74	1004		

7.11 THE CHARACTERISTIC FLUORESCENCE FACTOR, F

Note that we are going to neglect fluorescence by the continuum because it is only practical to evaluate by computer and its contribution is generally less than the other correction factors. Unfortunately, the continuum fluorescence should not always be neglected if one wants the most accurate answers. Its contribution is likely to be greatest when the matrix contains a large proportion of high Z elements and the element to be measured is middle or low Z; then the continuum intensity due to the matrix elements is appreciably stronger than that representative of the element to be determined. Fortunately in such cases, the atomic-number correction is likely to be larger than the continuum fluorescence so that the continuum fluorescence may still be considered approximately as a second-order term. All of this points to the advantage of using computer programs which can include the continuum fluorescence contribution.

The characteristic fluorescence term given below comes from the work of Castaing (63) and of Reed (70). The multiplier F in eq. 7.7 has the form $F = (1 + r_i)$ where

$$r_i = \sum_j P_{ij} C_j \left(1 - \frac{1}{J_i}\right)\left(\frac{U_j - 1}{U_i - 1}\right)^{1.67} \frac{\omega_j}{2} \frac{\mu_{ij}}{\mu_{Mj}}$$
$$\times \frac{A_i}{A_j}\left[\frac{\ln(1+u)}{u} + \frac{\ln(1+v)}{v}\right] \quad (7.13)$$

where P_{ij} is a constant having the value 1 for K radiation exciting K fluorescence or L radiation exciting L fluorescence, the value 4 for K radiation exciting L fluorescence, and the value $\frac{1}{4}$ for L radiation exciting K fluorescence. The terms U_j and U_i are V/V_j and V/V_i; A_i and A_j are the atomic weights of elements i and j; J_i is the jump factor for element i, tabulated in Appendix A.3; ω_j is the fluorescent yield for element j, tabulated in Appendix A.2; $u = \mu_{Mi} \csc \psi/\mu_{Mj}$; and $v = \sigma/\mu_{Mj}$ where σ was defined in Sect. 7.9.

Additional modification to F may be found in the work of Criss (73).

7.12 NUMERICAL EXAMPLE OF THE CORRECTION FACTOR EQUATION

As a simple example of a three-component system let us consider an alloy estimated to contain 30% iron, 30% nickel, 40% molybdenum. This is obviously not a practical alloy but is chosen to illustrate the

EXAMPLE PROBLEM

various correction factors. Let $V = 30$ kV and $\psi = 30$. We will calculate the intensity for iron.

First we will calculate the absorption factor A.
For the pure iron:

$$\chi_{100\text{Fe}} = \mu_{\text{FeFe}} \csc 30 = 76 \times 2 = 152$$

$$\sigma = \frac{4.5 \times 10^5}{(30 - 7.1)^{1.65}} = 2.57 \times 10^3$$

$$h = \frac{1.2 \times 55.8}{26^2} = 0.100$$

From eq. 7.8, $f(\chi_{100\text{Fe}}) = 0.94$
For the alloy:

$$\chi_M = \csc 30 \times (0.3\mu_{\text{FeFe}} + 0.3\mu_{\text{NiFe}} + 0.4\mu_{\text{MoFe}})$$
$$= 2 \times 159 = 318$$

$$h_M = \frac{1.2}{\Sigma}\left(C_i \frac{Z_i^2}{A_i}\right) = \frac{1.2}{15.01} = 0.080$$

$$f(\chi_M) = 0.880$$

Therefore $A = 0.880/0.940 = 0.936$

Next we will calculate the atomic number factor Z:
To enter Table 7.3 we need $1/U_{\text{Fe}}$ which is V_{Fe}/V, and so on.

$$1/U_{\text{Fe}} = 7.1/30 = 0.24; \quad 1/U_{\text{Ni}} = 0.28; \quad 1/U_{\text{Mo}} = 0.67$$

From Table 7.3 by interpretation, $R_{\text{Fe}} = 0.859; R_{\text{Ni}} = 0.854; R_{\text{Mo}} = 0.893$

$$R_M = 0.3 \times 0.859 + 0.3 \times 0.854 + 0.4 \times 0.893 = 0.871$$

Using eq. 7.12 or Figure 7.9 and eq. 7.11

$$\bar{S}_{\text{Fe}} = \frac{26}{55.8} \ln \frac{1.17(30 + 7.1)/2}{332} = 1.95$$

$$\bar{S}_{\text{Ni}} = \frac{28}{58.7} \ln \frac{1.17(30 + 8.3)/2}{362} = 1.96$$

$$\bar{S}_{\text{Mo}} = \frac{42}{96.0} \ln \frac{1.17(30 + 20)/2}{567} = 1.71$$

$$\bar{S}_M = 0.3 \times 1.95 + 0.3 \times 1.96 + 0.4 \times 1.71 = 1.86$$

Hence

$$Z = \frac{R_M \bar{S}_{\text{Fe}}}{R_{\text{Fe}} \bar{S}_M} = \left(\frac{0.871}{0.859}\right)\left(\frac{1.95}{1.86}\right) = 1.06$$

Now we are ready to calculate the fluorescence correction F. Only the NiK radiation will excite the iron strongly. The terms in eq. 7.13 have the following values:

$P_{\text{FeNi}} = 1$ because K radiation is exciting K fluorescence

$$C_{\text{Ni}} = 0.3$$

$$1 - \frac{1}{J_{\text{Fe}}} \approx 0.9$$

$$U_{\text{Ni}} - 1 = \left(\frac{30}{8.3}\right) - 1 = 2.62$$

$$U_{\text{Fe}} - 1 = \left(\frac{30}{7.1}\right) - 1 = 3.22$$

$$\left(\frac{U_{\text{Ni}} - 1}{U_{\text{Fe}} - 1}\right)^{1.67} = 0.706;$$

$$\omega_{\text{Ni}} = 0.375; \frac{\omega_{\text{Ni}}}{2} = 0.188$$

$$\mu_{\text{FeNi}} = 400$$

$$\mu_{\text{MNi}} = 0.3 \times 400 + 0.3 \times 59 + 0.4 \times 180 = 220$$

$$\frac{\mu_{\text{FeNi}}}{\mu_{\text{MNi}}} = 1.82$$

$$\frac{A_{\text{Fe}}}{A_{\text{Ni}}} = \frac{55.8}{58.7} = 0.95$$

$$\mu_{\text{MFe}} = 0.3 \times 76 + 0.3 \times 94 + 0.4 \times 270 = 159$$

$$u = 159 \times \frac{2}{220} = 1.45$$

$$\frac{\ln(1 + \mu)}{u} = \frac{0.895}{1.45} = 0.618$$

$$v = \frac{\sigma_{\text{Ni}}}{\mu_{\text{MNi}}} = \frac{2.9 \times 10^3}{220} = 13.2$$

$$\frac{\ln(1 + v)}{v} = \frac{2.65}{13.2} = 0.201$$

Thus

$$r_{Fe} = 1 \times 0.3 \times 0.9 \times 0.706 \times 0.188 \times 1.82 \times 0.95(0.618 + 0.201)$$
$$= 0.051$$

and $F = 1.051$

Substituting the values for Z, A, and F into eq. 7.7 we obtain

$$K_{Fe} = 0.3 \times 1.06 \times 0.936 \times 1.051 = 0.313$$

which means that the relative x-ray intensity of iron in the alloy is higher than the fractional composition because the increase due to atomic number effect and characteristic fluorescence outweighs the absorption effect.

7.13 COMPUTER PROGRAMS AND ITERATION

It is easy to see that the hand calculations become very tedious in practice and it is understandable that they have been computerized by many different workers. One of the computer programs most popular currently is the MAGIC program of Colby.* It allows the user to choose which expressions he will use for the Z, A, and F factors.

Each of the computer programs is designed to iterate the results automatically. Usually the measured relative intensities for the elements are used as the first approximation to composition (with or without normalization to 100%) and the computer program proceeds to find calculated intensities for such a composition. The calculated intensities are compared to the measured values and an adjustment made in the estimated composition. Then the program calculates a new set of intensities and again compares them with the measured values. This iteration proceeds until a satisfactory match is obtained between measured and calculated values. The estimate of composition for which the satisfactory match is obtained is taken as the best value for composition.

In the iteration approach there have been several suggestions (74, 75) for the best way to adjust the mismatch between calculated and measured intensities to obtain the next estimate of composition. One of the simplest approaches, and a usually satisfactory one, is illustrated in Figure 7.10 (73).

The calculated value of intensity for a given element is assumed to lie on a hyberbolic calibration curve which passes through the zero and

---

* Available from J. W. Colby, Bell Telephone Labs. Inc., 555 Union Blvd., Allentown, Pennsylvania, 18103.

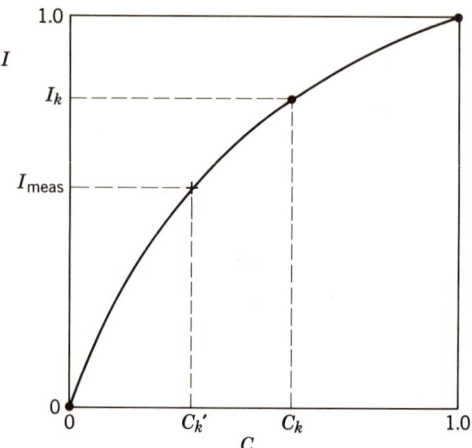

Fig. 7.10 The hyperbolic method for iteration. The first hyperbola is determined by the end points and calculated value of I_k corresponding to an assumed value of C_k. The next assumed value C'_k is the one corresponding to I_{meas}. In the new calculation if I_k does not match I_{meas} it determines a new hyperbola, and so on until I_{calc} matches I_{meas}.

100% points as well as the point C_k representing the previous assumed composition. The next estimate of composition is obtained by moving along the hyperbola to point C'_k which represents the measured intensities. With several elements in a sample, the hyperbolae change slightly with each iteration step but all of the mathematics for the analysis is contained in the computer program and the estimating is done automatically. A somewhat different adjustment procedure (74) favored by some workers uses the previous three values of calculated intensities to estimate the new approximation to composition. According to Reed (76) the Wegstein adjustment is more certain to seek the proper value but this author is not thoroughly convinced and finds the Criss (73) adjustment quite satisfactory and simpler to apply.

7.14 ERRORS DUE TO PARTICLE SIZE OR SHARP BOUNDARIES

The expressions considered in the previous sections of this chapter are based on local homogeneity at least within the effective x-ray source size. This situation will not prevail near sharp boundaries such as the edges of diffusion zones or for precipitates smaller than a few microns. In such cases the x-ray intensity cannot easily be related to composition. For instance, Figure 7.11 is a micrograph of an undiffused iron–nickel

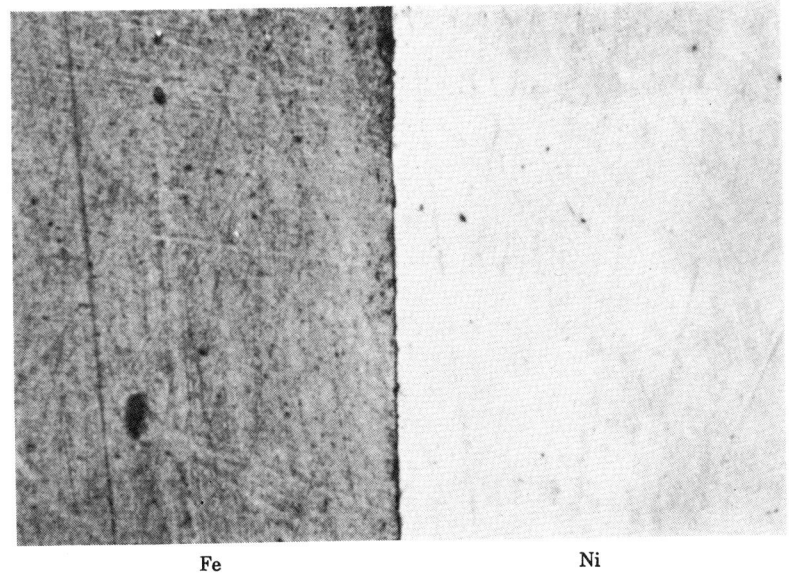

Fig. 7.11 Micrograph of the sharp boundry between iron and nickel is an undiffused couple.

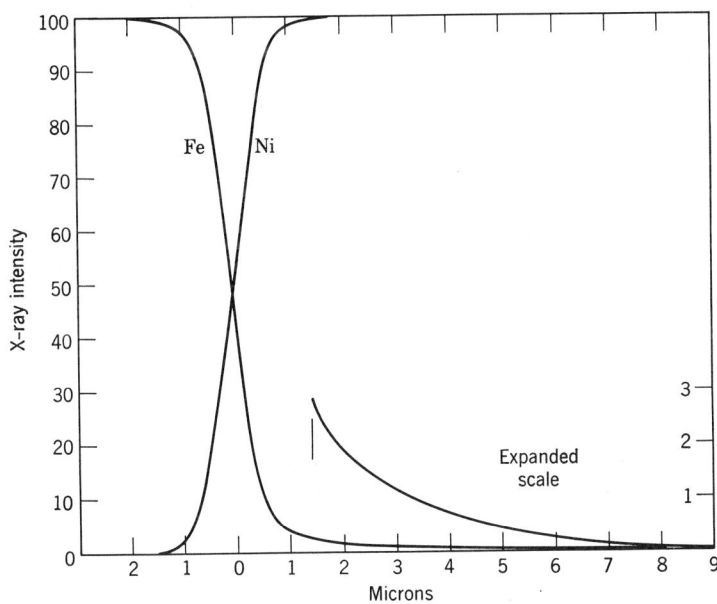

Fig. 7.12 X-ray intensity for FeK_α and NiK_α across the boundary of Fig. 9.2. Secondary fluorescence of FeK_α by nickel radiation causes excitation of FeK_α intensity even when the electron beam strikes several microns away from the iron.

couple and Figure 7.12 shows the iron and nickel intensities as the electron beam is scanned across the junction. When the beam strikes on the iron side of the couple, the intensity represents the proper 100% Fe and 0% Ni except within 2μ of the boundary. But when the beam is on the nickel side of the couple, the nickel radiation generated penetrates through to the iron side, excites iron by secondary fluorescence, and results in an iron reading although there is no iron present. The effect persists according to Figure 7.12 until the beam has moved some $7-8\mu$ away from the boundary. This is a particularly bad case because of the strong secondary fluorescence; for many cases one can expect accurate quantitative analysis except within $2-3\mu$ from a sharp edge.

A similar situation exists for depth of penetration when one examines thin layers or precipitates which are too thin to absorb the x-rays generated. Figure 7.13 shows the fractional contribution to emitted intensity GaK_α in GaAs as a function of depth below the surface. The primary generation of GaK_α is limited to about 2μ but secondary fluorescence of GaK_α by AsK_α persists to depths of $7-8\mu$.

As yet there are no mathematical corrections which take account of local inhomogeneities in performing quantitative analysis. Such expressions would undoubtedly be very complex. The best approach would

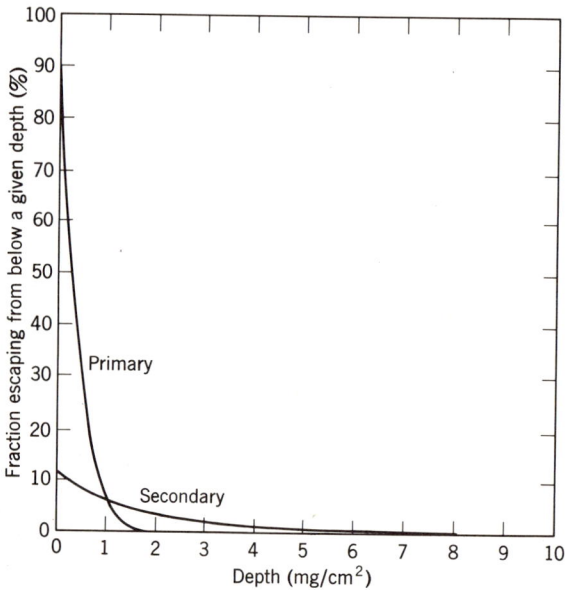

Fig. 7.13 Monte Carlo calculation of the distribution in depth for primary excitation and secondary fluorescence excitation of GaK_α in GaAs.

appear to be a Monte Carlo calculation incorporating some preliminary estimate of size of zones and distribution of elements based on an ordinary semiquantitative analysis across the inhomogeneous region. This would be prohibitively expensive until computer speeds have increased at least an order of magnitude beyond their 1969 level. Extraction replicas discussed in Section 6.5 allow precipitates to be examined without interference from the matrix but there is no correspondingly easy way to examine the matrix without interference from the precipitates. This can be a serious problem when one is trying, for instance, to study matrix depletion in chromium near chromium-rich precipitates.

CHAPTER

8

ADVANCED COMPUTER METHODS FOR QUANTITATIVE ANALYSIS

In the previous chapter a number of routine methods for quantitative analysis were discussed. All of them were based on approximations required for pencil-and-paper evaluation although computer programs are often used to speed the calculations. With the availability of higher-speed computers in about 1965, it became feasible to consider expressions for the physical processes of x-ray generation with less approximations than were necessary for the pencil-and-paper equations. These expressions have their own approximations, of course, but in general they are more rigorous and less phenomonological.

One of the most powerful second-generation approaches is the electron transport program of D. B. Brown (77, 78) and it will be discussed in some detail in this chapter although it is not in common use. The computer running time is 10 to 100 times longer than for the computerized correction-factor approaches of Chapter 7 (but still only 34 sec on a CDC 3800 computer) and it would not be practical, at present, to use the transport method for all iteration stages. However a combination of a fast, approximate approach for the first two or three iterations (see Section 7.13 for a discussion of iterations) and the transport approach for the final one or two iterations does seem practical and is, in fact in operation. The next higher stage of sophistication appears to be Monte Carlo calculations combined with analytic expressions wherever possible. With the computer speeds of 1969, the Monte Carlo approach would require more than ten times longer than the electron transport approach and is prohibitively expensive except for test cases to determine subtle effects in critical instances. It will not be discussed in detail because of the lack of programs available but the reader is advised that it may begin to come into favor by about 1975–1980.

It should be noted in this chapter that the intent is not to make the analyst an expert in the mathematics necessary to comprehend all that is done in the electron-transport computer program described. Rather it is intended only to point out some of the general concepts employed so that he will appreciate what the computer program is doing for him. The point is, *it may be better for the analyst to use a sound mathematical procedure which he does not comprehend than a simplistic one which he can understand but which lacks efficacy.*

8.1 THE ELECTRON-TRANSPORT CONCEPT

The electron-transport equations calculate the number of electrons at each depth below the surface of the sample, their energy, and their direction. Knowing these parameters one can calculate the number of characteristic x-rays generated using the K or L ionization cross-sections. The electron transport method evaluates the complete electron generation for the actual matrix rather than separating it into $f(\chi)$ terms which take account of the absorption factors for the actual matrix and $\phi(\rho z)$ terms (atomic-number correction) which are needed to account for the effect of atomic number on the generation of x-rays (see Section 7.10). Secondary fluorescence due to characteristic radiation and the continuum are not part of the electron transport mathematics but they are included in the transport computer program. It seems likely that if one seeks accuracy of the order of 2% relative, then the differences between the electron transport method and the approximate corrections of Chapter 7 are important because that is about the limit of the approximation corrections.

It is not appropriate here to repeat in detail all of the mathematics of the transport approach so only a brief explanation is given. Basically, the electron transport program attempts to solve what is called the Boltzman equation for the position and energy of the electrons as they move into the specimen. The Boltzman equation says that if the position and energy of a group of electrons are known at one point along their path, the position and energy may be determined at a later time by considering two terms: one describes how electrons are scattered and the other describes how they lose energy. To solve the problem, all of the incident electrons of energy E_0 are considered to strike the sample in a group at zero time. Their later position or depth, z, below the surface; their angle, ϕ, with respect to the incident beam; and the path length traveled, s, must be expressed as a multivariate distribution function $f(s, z, \phi)$ because they are no longer traveling in a group.

It is assumed that the energy, E, for any given electron is simply related to its path length, s, traveled and that the average energy loss per unit path length, dE/ds, at any point on the path, can be expressed by the theory of Bethe (79) with some modifications (80, 81). In particular, the form used is

$$\frac{dE}{ds} = 7.88 \times 10^4 \left[\frac{2E + 511}{511}\right] \frac{\rho}{E} \sum_i \left[C_i \left(\frac{Z}{A}\right)_i \left(0.154 + \ln \frac{E}{I_i}\right)\right] \quad (8.1)$$

where E is in keV, s is the distance traveled in cm, ρ is density, C_i and $(Z/A)_i$ are the concentration and ratio of atomic number to atomic weight for element C, and I_i is the "average excitation energy" for element i (the averaging is over the various excitations in the K, L, M shell).

The scattering is treated according to the paper of Bethe, Rose, and Smith (82). Following their treatment we are concerned with the scattering angle, ϕ, and the cross-section for scattering, $\sigma(\phi)$. The cross-section is expressed in terms of a mean free path, $\lambda(s)$, for electrons of a given energy, E, that is, for a given path length, s.

$$\frac{1}{\lambda(s)} = \pi N \int_0^\pi \sigma(\phi)[1 + \cos \phi] \sin \phi \, d\phi$$

Then the transport equation for the distribution of electrons at distance Z below the surface, traveling at angle ϕ with respect to the incident beam, and having traversed a path length s, is written in terms of the variation with path length

$$\frac{\partial}{\partial s}[\sin \phi f(z, \phi, s)] = -\cos \phi \frac{\partial}{\partial z}[\sin \phi, f(z, \phi, s)]$$

$$+ \frac{1}{\lambda(s)} \frac{\partial}{\partial \phi}\left[\sin \phi \frac{\partial}{\partial \phi} f(z, \phi, s)\right] \quad (8.4)$$

It is not possible to solve the transport equation explicitly but it can be done with a computer by using what are called difference equations. To understand how this is done, let us imagine a series of cards, each one representing a distance, s, along the path. On each card the

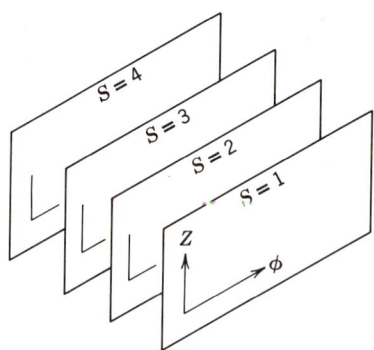

Fig. 8.1 An imaginary series of cards for different electron path lengths, S_1, S_2, etc. For a given distance S the card contains information on the number of electrons at depth Z and traveling at angle ϕ with respect to the original beam.

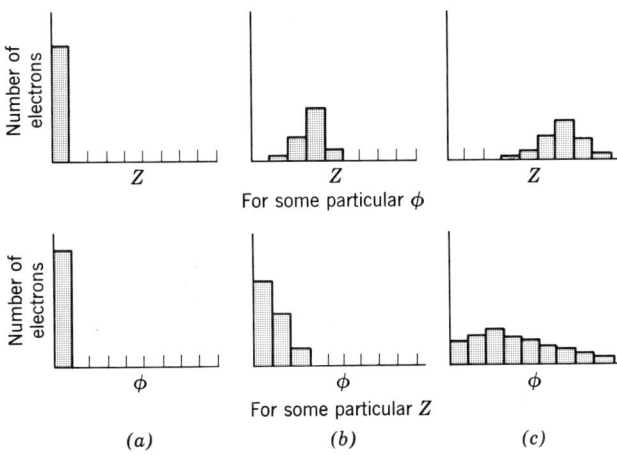

Fig. 8.2 The kind of information about Z and ϕ. On the first card corresponding to the incident beam striking the surface all the electrons will be at zero depth and traveling at $\phi = 0$. For the second card, after each electron has traveled some distance in the specimen, for each value of ϕ there will be a distribution in Z and for each value of Z there will be a distribution in ϕ. For some later card the distribution will be more smeared out for both Z and ϕ. (a) First card; (b) Second card; (c) Later card.

distribution in direction, ϕ, and depth below the surface, z, are given as shown in Figure 8.1. For the first card both z and ϕ are single valued if we assume that the electrons are in a parallel beam and all strike the sample at one time. That is, the distributions would be as in Figure 8.2a.

To find the number of electrons at each specific value of z and ϕ on the next card we consider those electrons in the range $\phi \pm \Delta\phi$ and z to $z = \Delta z$ on the previous card. That is, we find the distribution at $s + \Delta s$ from the difference equation

$$\sin \phi f(s + \Delta s, z, \phi) = \sin \phi f(s, z, \phi) + P(s) + S(s)$$

where

$$P(s) = -\left(\frac{\Delta s}{\Delta z}\right) \cos \phi \sin \phi \, [f(s, z, \phi) - f(s, z - \Delta z, \phi)]$$

and

$$S(s) = -\frac{1}{\lambda(s)} \frac{\Delta s}{(\Delta\phi)^2} \left\{ \left[\sin\left(\phi + \frac{\Delta\phi}{2}\right)\right][f(s, z, \phi + \Delta\phi) - f(s, z, \phi)] \right.$$
$$\left. + \left[\sin\left(\phi - \frac{\Delta\phi}{2}\right)\right][f(s, z, \phi - \Delta\phi) - f(s, z, \phi)] \right\}$$

This would give distributions something like those shown in Figure 8.2b where there would be a distribution in z for each value of ϕ and vice versa. For some later card the distributions would appear as in Figure 8.2c, again with a distribution in z for each value of ϕ and vice versa. It should be noted that the distribution in z has moved to a greater average depth but the distribution in ϕ has merely spread out to be more uniform in angle.

The generation of x-rays at each depth, z, is calculated by multiplying the number of electrons at that depth interval on each of the cards by three factors: (*1*) the ionization cross-section for the element and line of interest; (*2*) the assumed concentration of that element; and (*3*) the fluorescent yield (83) for that element and line. Absorption of the emerging radiation is treated by the usual exponential absorption law for the assumed matrix composition. Actually there are really no cards but all of the equivalent calculations are carried out in the computer.

Several modifications to the simple program outlined above have been included by Brown (78) to make the results more accurate. For the initial part of the path the scattering is treated in a somewhat more detailed manner because large-angle scattering is relatively more important at that stage of transport than it is later when the electrons have been spread out in angle. For the final stages of the path where the electron directions are sufficiently random, the electron transport is treated as a diffusion process which shortens the computer time required for the remainder of the calculation. Other recent improvements (78) include a modification of the electron scattering expression to account for the effect of neighboring atoms on the scattering potential distribution around the atoms and a modification to account for straggling, that is, to account for those electrons which have lost more energy than given by the average loss in eq. 8.1.

8.2 RAMIFICATIONS AND LIMITATIONS OF THE TRANSPORT APPROACH

It has been demonstrated that the transport method already gives answers which are equally as satisfactory as the other calculation methods (84) but this is hardly sufficient to warrant using a more expensive calculation method. The feature which makes transport calculations more attractive is that they do not require any arbitrary fitting factors such as the J/Z values (70) which have been adjusted to match presently available data but which may not be satisfactory for all future analysis problems. The main uncertainty in the transport equation (aside from the errors in absorption coefficients and fluorescent yields which plague all of the quantitative methods at present) is the stopping-power expression for electrons and the effects of electron-shell structure in the atoms on the stopping power. This is a fundamental problem in physics which will eventually be solved and will place the transport method in a very favorable position, hopefully in the near future.

8.3 MONTE CARLO METHODS

Limited Monte Carlo calculations have already been used to calculate the electron trajectories in a sample (85–87) and to calculate the contribution from secondary fluorescence as was shown in Figure 7.13 (88). But these were limited problems and do not represent the extensiveness of a complete Monte Carlo treatment of each incident electron in complex samples and the resulting electron paths and generation of x-rays. Even with the fastest present computer speeds such a problem would be prohibitively expensive as an analytical tool. However it seems reasonable to expect that a few complete Monte Carlo calculations will be attempted for critical types of composition as a test of the efficacy of the other methods. It only remains for some laboratory to write the program and devote sufficient time and money to test it. Monte Carlo programs of equal complexity are already being used in γ-ray transport (89) and other areas where the answers are sufficiently valuable to warrant the expense.

CHAPTER

9

RELATED INSTRUMENTATION AND TECHNIQUES

There are three kinds of instruments related to the electron probe: (*1*) instruments which have a similar purpose, namely the measurement of local variation in composition such as ion probe microanalyzers or electron spectrometers; (*2*) instruments or techniques which measure related properties of materials such as scanning electron microscopes, electron and x-ray microdiffraction, divergent beam techniques, and x-ray microscopy; and (*3*) instruments combining electron probes with other techniques such as electron microscopy or electron spectroscopy.

It is not the purpose here to go into detailed discussion of these instruments but only to show their general purpose and approach and their relation to electron probe microanalysis.

9.1 THE ION PROBE MICROANALYZER

One type of ion probe was first developed by Castaing and Slodzian (90) in 1962 and is presently manufactured by CAMECA (91). It bombards an area of a few hundred microns on the surface of a sample with argon or oxygen ions causing emission of secondary ions from the elements in the sample. Figure 9.1 shows the instrument schematically. The secondary ions are passed through a magnetic mass spectrometer to separate and select an individual ion species and through an electron mirror to select a small energy range for that particular species (to optimize spatial resolution). The final image is formed by an ion-to-electron converter with the electrons striking a fluorescent screen. The electron image on the screen may be photographed or viewed visually or the electron current in a 5 μm area may be measured by a scintillation-photomultiplier placed behind a small aperture. The measurement of current is related to the concentration of the element at that point on the sample but quantitative analysis is far more difficult than in electron probe analysis because of the complex intensity interrelationships resulting from surface contamination as well as more complex matrix effects for emitted ions compared to characteristic x-rays. The surface of the specimen is eroded by ion emission at a rate of about 50 Å/sec; thus one obtains depth resolution with time and may photo-

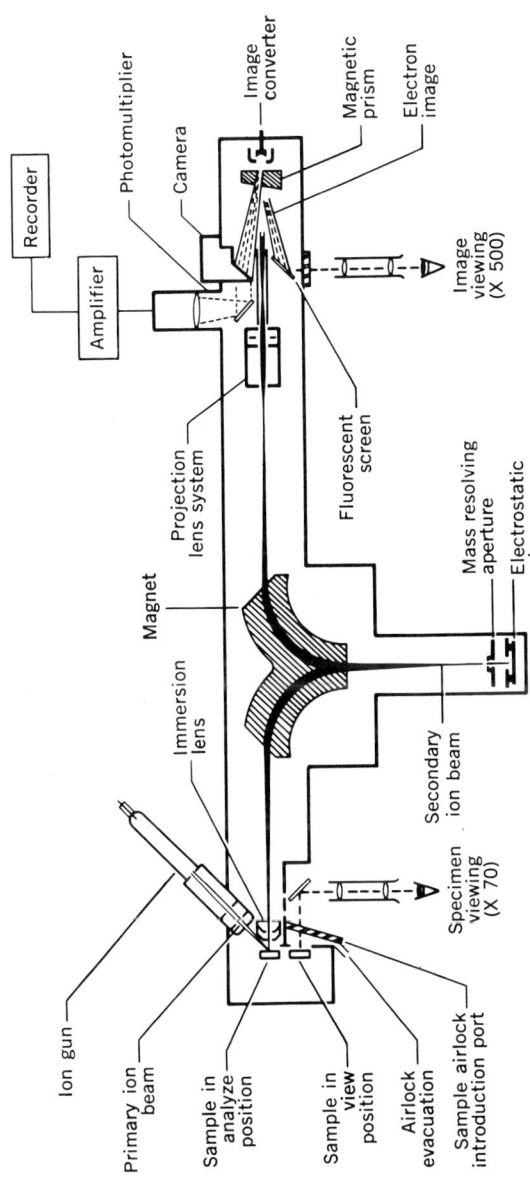

Fig. 9.1 The CAMECA imaging type of ion probe. (Reproduced from Research/Development, Feb. 1970, p. 36, © 1970 by F. D. Thompson Publications, Inc.)

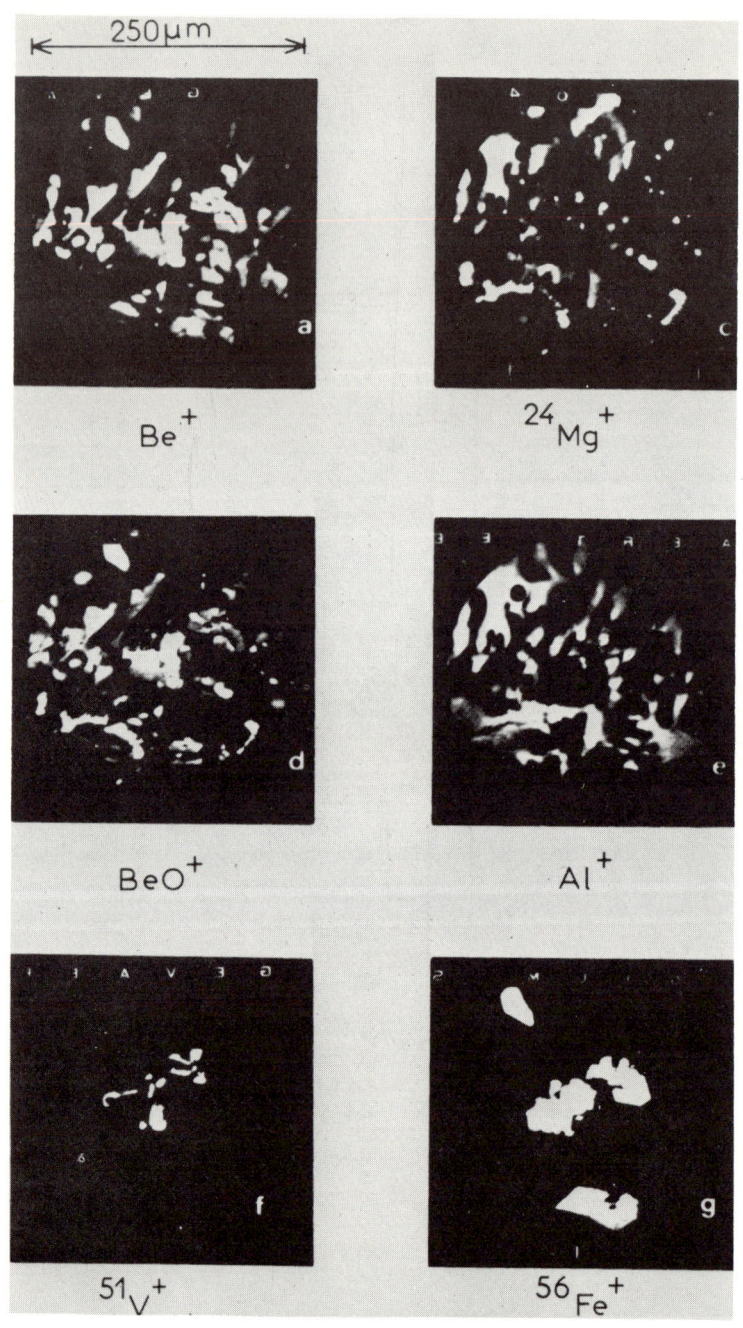

Fig. 9.2 Photographs taken with the CAMECA ion probe showing the distribution of different elements. (Courtesy CAMECA, Paris, France.)

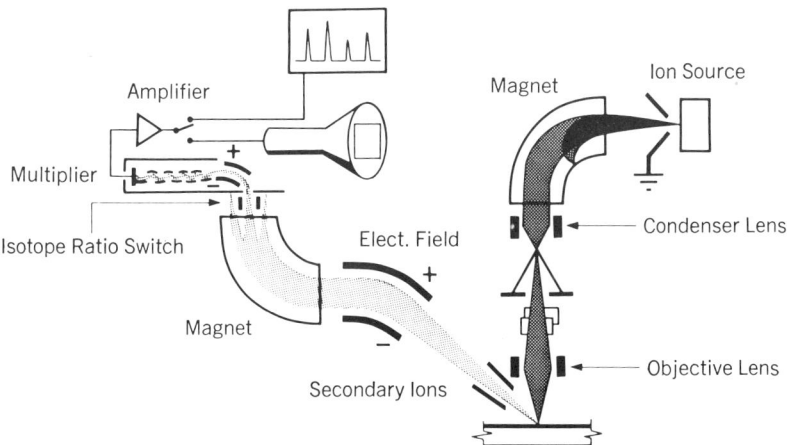

Fig. 9.3 The Applied Research Laboratories scanning ion probe. (Courtesy ARL, Sunland, California.)

graph successive layers down into the specimen as the surface is removed. Of course this technique is destructive of the surface unlike electron probe analysis. Figure 9.2 shows ion micrographs of several elements in a cast aluminum alloy taken with the CAMECA ion probe.

A second approach to ion probe microanalysis was developed by Applied Research Labs (92). In their instrument, Figure 9.3, the incident ion beam is focused to a 1-2 μm diameter so that only a local area of the sample is bombarded and hence the emitted ions come from that local area. Again a magnetic analyzer is used to select the ion species (including the individual isotope) which is passed through an aperture into a scintillation-photomultiplier detector. Scanning as in the electron probe allows an area of the surface to be examined.

Both types of ion probes can use either positive or negative ions and can detect the lowest-atomic-number elements at trace concentrations down to tens of parts per billion. There is no continuum background intensity as there is for the x-ray spectrum in electron probe analysis.

9.2 ELECTRON SPECTROMETRY

Electron spectrometry (93, 94) using either photoelectrons or Auger electrons already competes with x-ray fluorescence for measuring low Z elements in bulk samples and is likely to compete with electron probe analysis of local areas. Figure 9.4 shows an electron spectrometer schematically. Electrons emitted from the sample pass between cylindri-

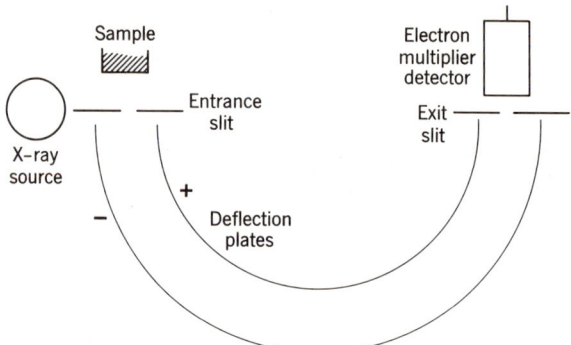

Fig. 9.4 An electrostatic electron spectrometer.

cal or spherical electrostatic focusing plates; the voltage on the plates determines the energy of those electrons which will be focused and pass out through the exit slit. If the plate voltage is varied with time the electron energy varies likewise and an electron energy spectrum results.

In photoelectron spectrometry electrons or monochromatic incident x-rays are absorbed by K or L shell ionization of atoms in the sample. The photoelectrons removed by the ionization process have a characteristic energy which depends on the binding energy and the energy of the incident quantum. If the incident quanta are x-ray photons the photoelectric peaks will be fairly sharp but if they are electrons the peaks will only be sharp on the high energy side because an incident electron may lose any amount of energy up to its maximum value in ionizing the atom.

In Auger electron spectrometry we neglect the photoelectrons which are removed during K or L shell ionization of the atoms. Instead, we use the Auger electrons which are emitted when ionized atoms decay to the ground state by a radiationless transition. This process is often explained by saying that the characteristic x-ray photon which should be emitted is reabsorbed by ejecting one of the other electrons in the atom. That electron is an Auger electron and has an energy equal to the characteristic x-ray energy which would have been emitted minus the binding energy of the ejected electron. Thus the Auger peaks are sharp no matter what the source of initial ionization. Figure 9.5 shows both photoelectron and Auger peaks for magnesium excited by AlK radiation.

Photoelectron peaks are more intense than Auger peaks because one photoelectron results for each ionization whereas an Auger electron results only when a characteristic x-ray is not emitted; the low fluorescence

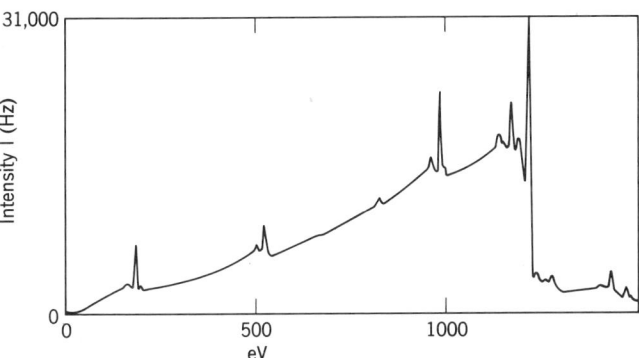

Fig. 9.5 Auger spectrum (and photoelectron peak) from magnesium excited by AlK radiation.

yield of less than 1% however, for elements like carbon or oxygen, means that more than 99% of the ionizations do result in Auger electrons but the intensity is distributed in several peaks. Usually electron spectrometers operate by differentiating the electron intensity signal as the focusing-plate voltage is varied. This means that the photoelectron or Auger peaks are enhanced above the generally strong but nearly constant background.

9.3 SCANNING ELECTRON MICROSCOPY

Scanning images with specimen current or backscattered electrons have been discussed throughout the volume of this book but always for instruments which are designed to allow characteristic x-rays to be measured easily. As explained in Chapter 3, the design of the second electron lens is limited by the x-ray optics requirements. Currently there are a number of instruments which do not make provision for measuring characteristic x-rays (at least not with x-ray spectrometers) and therefore do not have such limitations. Such instruments are called scanning electron microscopes and the beam size may be reduced to about 200 Å instead of 0.5 to 1 μm as usually employed for electron probes. Thus the spatial resolution is improved (95) as shown in Figure 9.6. Often the scanning electron microscopes are used to examine very rough surfaces (96) as shown in Figure 9.7 which would not be at all suitable for electron probe analysis. It should be noted that scanning electron microscopes can, and often do, incorporate an energy dispersion channel (Section 5.4) so that some information can be obtained on composition as well as surface configuration.

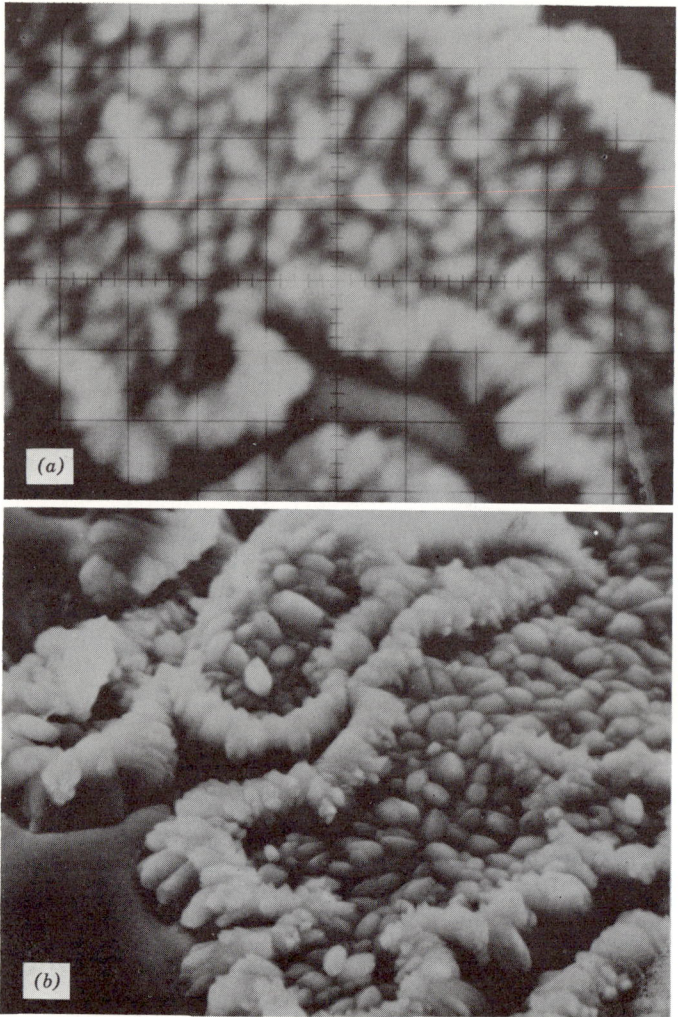

Fig. 9.6 Comparison of resolution with standard electron probe (a) and scanning electron microscope (b). (Courtesy E. J. Brooks, Naval Research Laboratories.)

9.4 ELECTRON AND X-RAY DIFFRACTION

A focused electron beam may be used directly for transmission or reflection electron diffraction (97) as shown in Figure 9.8. The purpose here, as with all diffraction, is to study crystal structure rather than chemical composition (electron probe) or physical configuration (scanning electron microscopy). The diffraction rings correspond to inter-

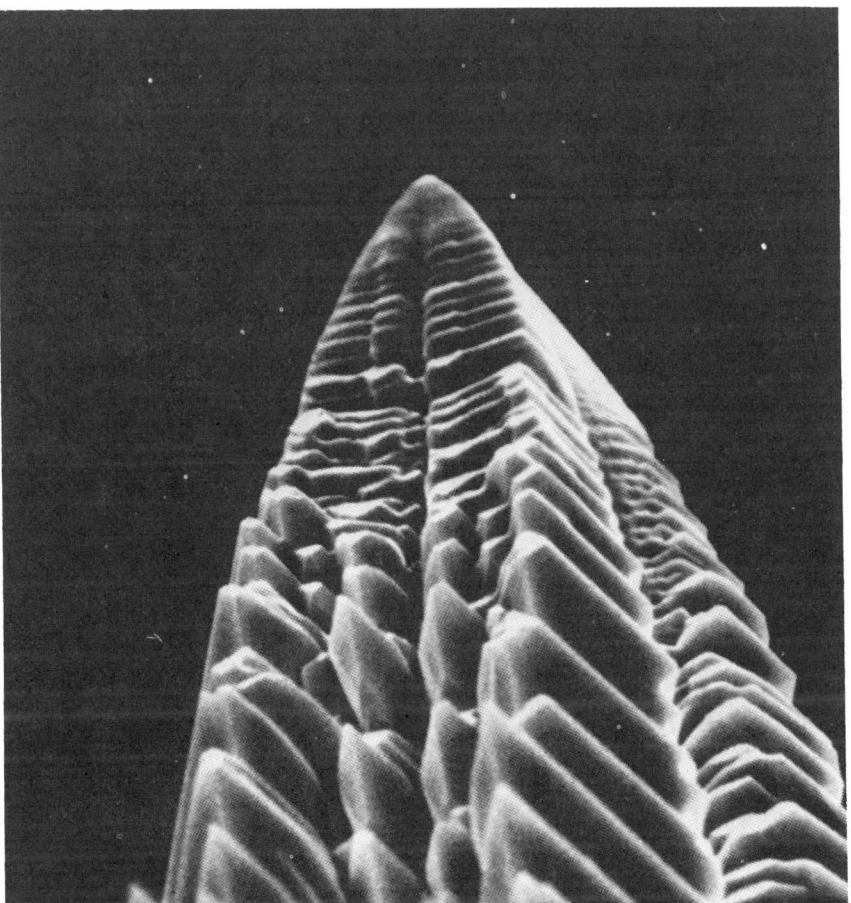

Fig. 9.7 Surface of tungsten trioxide crystal showing growth on crystallographic planes, 750×. (Courtesy S. Kimoto and J. C. Russ, *Amer. Sci.*, **57**, 1969.)

planar spacings in the crystals. Electron diffraction is usually not convenient in standard electron probes but is easily accomplished in combination electron probe–electron microscope instruments (see Section 9.7).

Another kind of electron diffraction called LEED (98) low-energy electron diffraction) employs a beam of a few tens of electron volts energy and a bulk specimen. The interaction of these low energy electrons with the specimen surface gives valuable information on monoatomic layers and surface energy states but the patterns are more difficult to

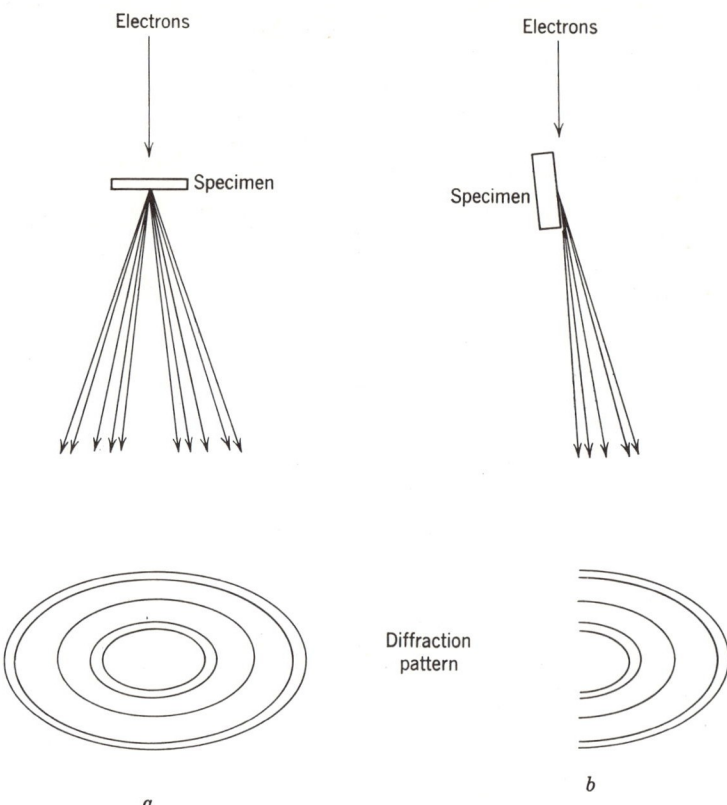

Diffraction pattern

a
b

interpret than ordinary diffraction patterns and the electron optics cannot be accommodated with the higher-energy electron optics of the electron probe.

For x-ray microdiffraction it is necessary to introduce a target and collimator system as shown in Figure 9.9. High specific loading of the target is required in order to obtain sufficient intensity in collimated x-ray beams on the order of 10-μm diameter. It can only be obtained by using a small-diameter electron beam as shown in Table 9.1. A single-lens electron optics system is adequate for x-ray microdiffraction and one can also do electron probe analysis of 3 to 5 μm areas if the x-ray target is replaced by the specimen. For x-ray microdiffraction it is desirable to move the target slowly during the course of the exposure in order to reduce contamination effects and keep the emitted x-ray intensity high. Exposure times vary between a few minutes and a few hours depending on the type of specimen.

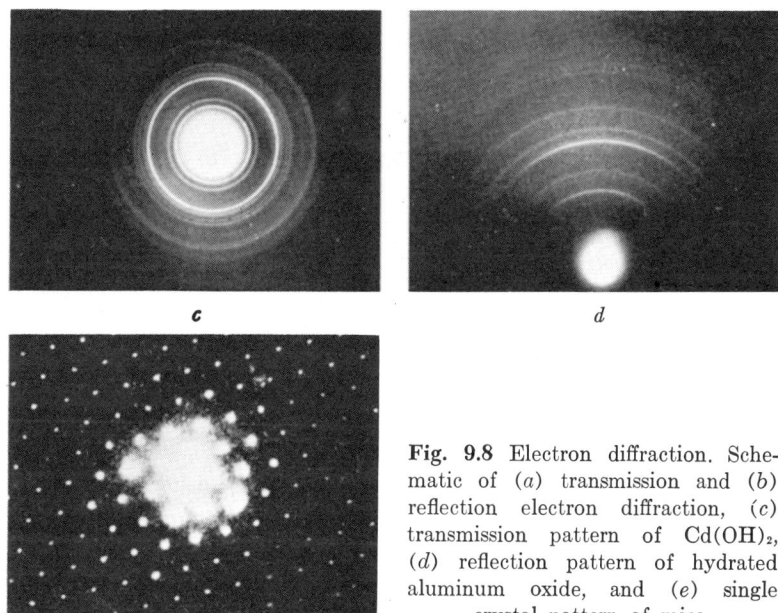

Fig. 9.8 Electron diffraction. Schematic of (a) transmission and (b) reflection electron diffraction, (c) transmission pattern of Cd(OH)$_2$, (d) reflection pattern of hydrated aluminum oxide, and (e) single crystal pattern of mica.

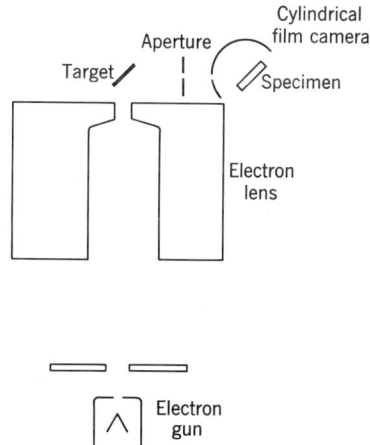

Fig. 9.9 Schematic diagram of electron optics used to generate small focal spot to be for x-ray microdiffraction.

Table 9.1. Approximate Specific Loading for Electron Beam Instruments (30 keV)

Instrument	Spot area (cm^2)	Current (A)	W/cm^2
Electron probe	10^{-8}	3×10^{-8}	9×10^4
Microdiffraction	10^{-6}	6×10^{-7}	1.8×10^4
Microfocus tube[a]	10^{-4}	10^{-4}	3×10^4
x-Ray tube[a]	10^{-1}	2×10^2	6×10^3

[a] Water-cooled target.

One advantage of x-ray microdiffraction in conjunction with electron probe analysis is that final identification of precipitates or inclusions may depend on knowledge of crystal structure as well as chemical composition. For instance, the metallic carbides $M_{23}C_6$ and M_4C cannot be distinguished by electron probe analysis alone because the difference in composition is beyond the limit of accuracy of quantitative analysis, but microdiffraction distinguishes the two places easily by their difference in crystal structure. In diffusion zones it is also of interest to measure variation of lattice parameter or orientation of phases.

9.5 DIVERGENT-BEAM PATTERNS

Another use for the electron beam is production of divergent-beam Kikuchi (99) or Kossel (100) patterns for study of crystal imperfections in single crystals or accurate measurement of lattice constants and wavelengths. The simplified principle of the Kikuchi patterns is illustrated in Figure 9.10a. Some of the electrons entering a crystal will diverge because of interaction with bound electrons in the lattice. Some of the diverging electrons will be traveling in the proper direction to satisfy the Bragg cones of diffraction from the planes shown in Figure 9.10a. The diffracted electrons are shown as solid arrows, and where the cones of diffraction strike the film dark arcs will appear. The dotted lines show the paths that would have been taken by the electrons had they not been diffracted. At the intersections of the dotted lines with the film there will be a reduction in background intensity that appears as a light arc.

Figure 9.10b is a reproduction of the Kikuchi pattern of fluorite taken from Thompson and Cochrane (97). In Kikuchi's original work with cleaved mica he found that a thickness of about 10^{-6} cm gave a pattern of light and dark arcs. Thus the simplified explanation given above

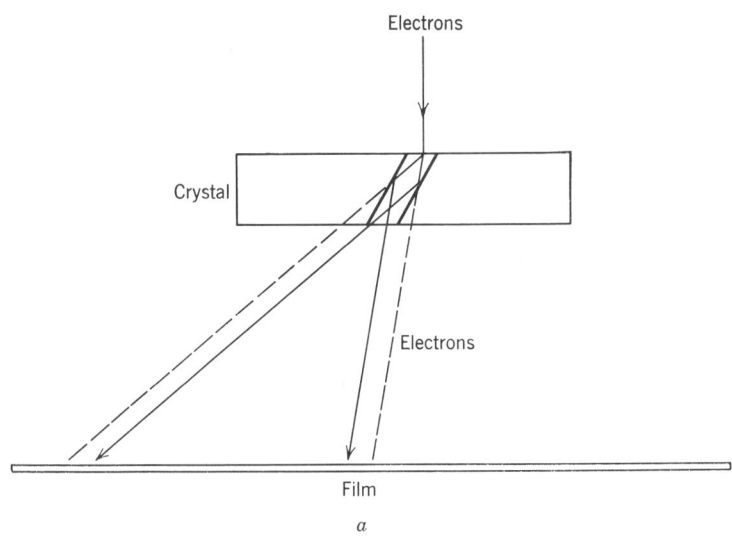

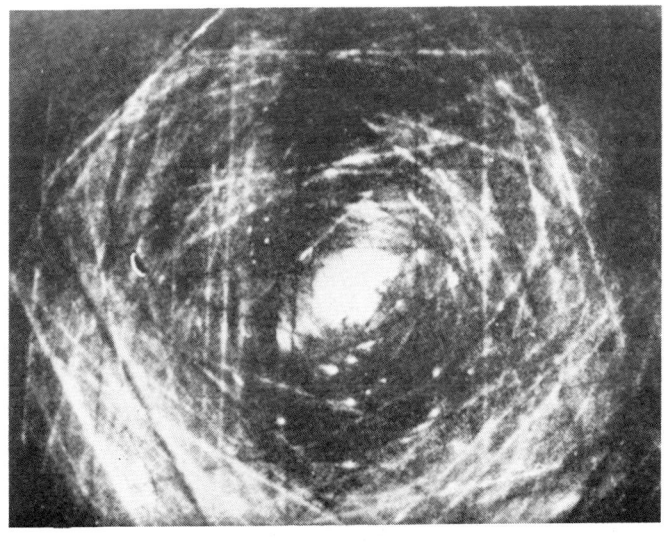

Fig. 9.10 (*a*) Schematic diagram of Kikuchi pattern formation; solid arrows represent dark lines on film; dotted lines represent corresponding light lines on film. (*b*) Kikucki lines from fluorite (97).

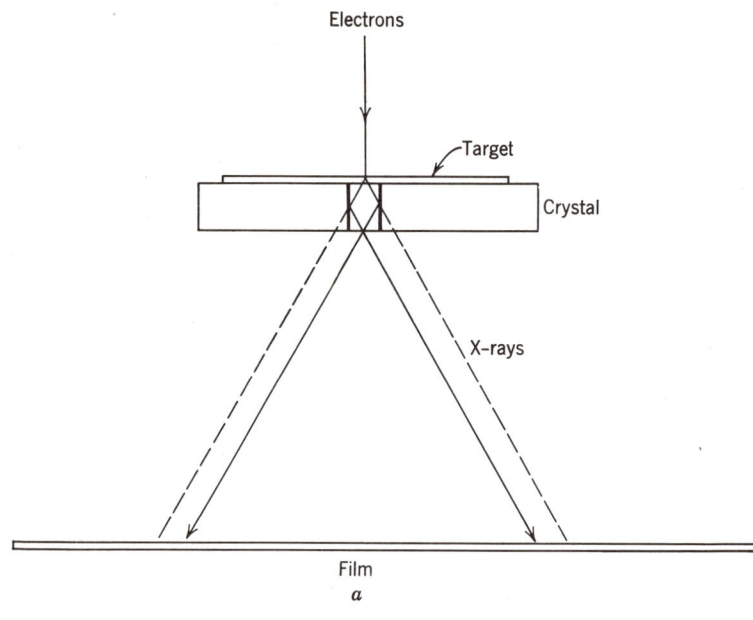
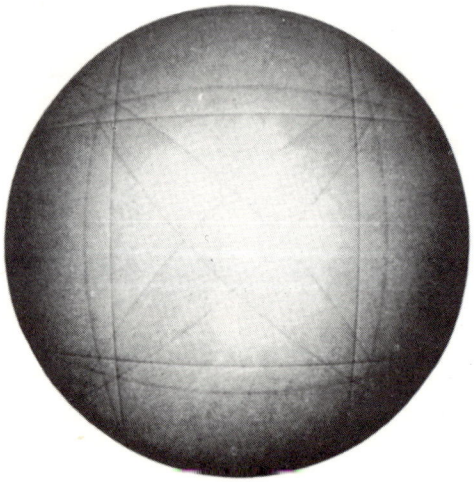

Fig. 9.11 (*a*) Schematic diagram of Kossel line pattern formation; solid arrows represent dark lines on film; dotted lines represent corresponding light lines on film. (*b*) Kossel line pattern of nickel single crystal. (Reprinted by courtesy of R. E. Ogilvie, Massachusetts Institute of Technology.)

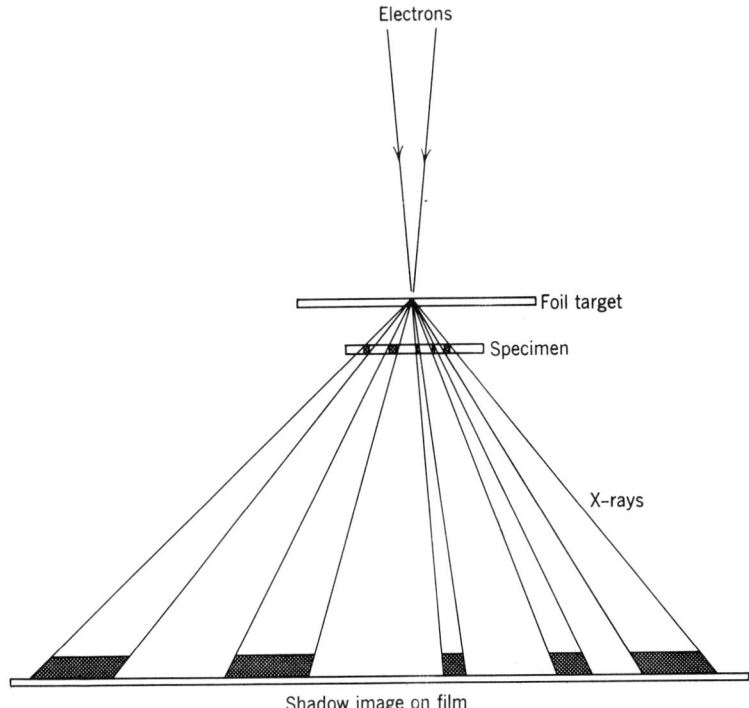

Fig. 9.12 Schematic diagram of x-ray projection microradiography. Regions of high density cast shadows on film.

is not sufficient to elucidate the phenomenon. For a theoretical discussion the reader is referred to von Laue (101) or Pinsker (102).

The analogous x-ray technique of Kossell as developed further by Lonsdale (103) is shown in Figure 9.11a. Electrons first strike a thin target and generate diverging characteristic x-rays. The target material is chosen to give x-rays of the desired wavelength. Divergent radiation strikes the crystal and satisfies the Bragg cones of diffraction as shown. Figure 9.11b is a Kossel line pattern of nickel from the work of Ogilvie and Moll and shows the light and dark arcs expected. They followed Kossel's original technique and did not use an intermediate target, but allowed the divergent x-rays to be generated in the crystal itself. This limits the crystals that can be examined to those substances that have suitable characteristic x-ray wavelengths.

The divergent electron or x-ray methods may be used to study imperfections in crystals because any change in the orientation of the planes will show up as distortion in the arcs. Also, from the distance between

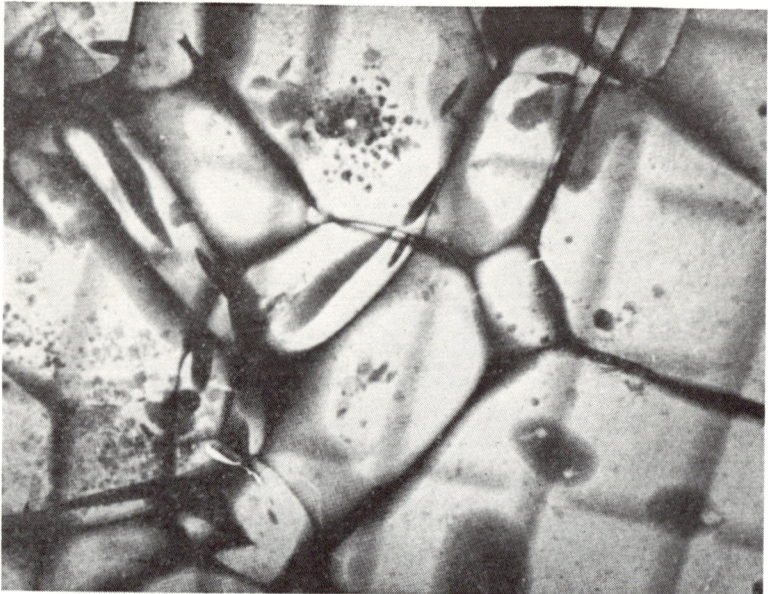

Fig. 9.13 Projection radiograph of aluminum–5% tin showing tin (dark) collected at the grain boundaries (104).

intersections of various arcs, very accurate measurements of interplanar spacing or x-ray wavelength may be obtained.

9.6 X-RAY MICROSCOPY

An excellent book by Cosslett and Nixon, *X-Ray Microscopy* (104), describes the use of electron beams for projection x-ray microscopy. Figure 9.12 shows the arrangement. Electrons strike a thin foil target and generate x-rays. The characteristic radiation from the target elements is the predominant wavelength. When a specimen is placed close to the target and a photographic film further away, the projection radiograph shows details of the specimen components because of variation in x-ray absorption. Figure 9.13 is a radiograph of aluminum–5% tin and has a resolution of about 1 μm. Microradiography is especially valuable in biology because living specimens may be observed. It is also very valuable in metallurgy and mineralogy, where filtered x-rays may be selected to show slight changes in composition of different phases.

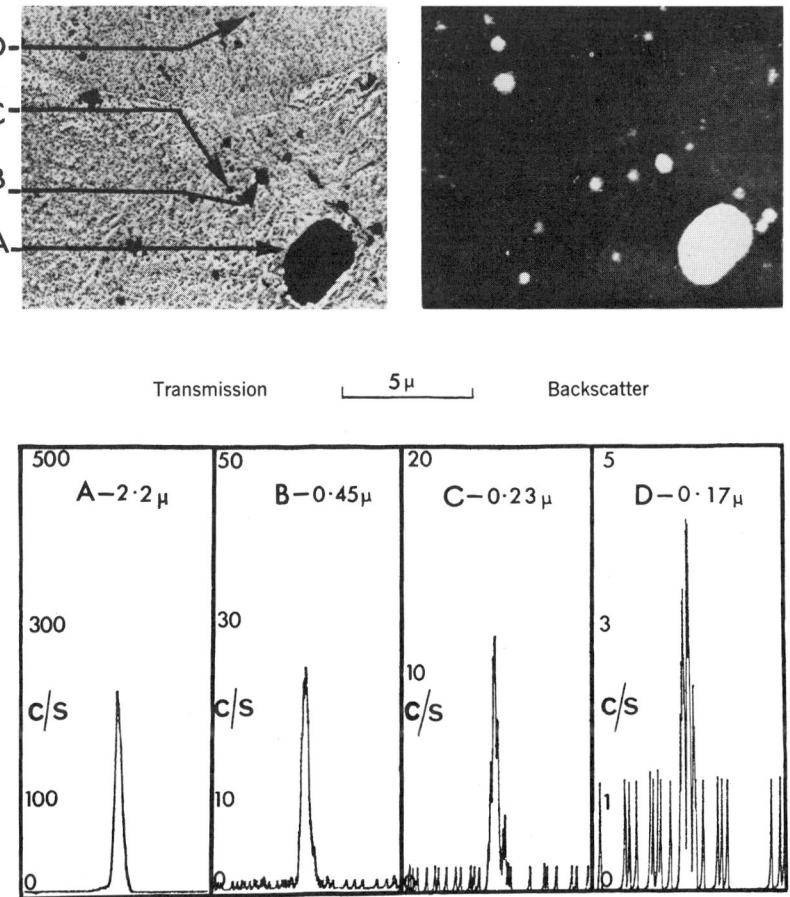

Fig. 9.14 The use of a combination electron probe–electron microscope for selecting small particles. The x-ray lines are Ti$K\alpha$ from the individual precipitates labeled. (Courtesy P. Duncumb, *The Electron Microprobe,* Wiley, New York, 1966, p. 490).

9.7 COMBINED ELECTRON PROBE–ELECTRON MICROSCOPE

Duncumb (105) was the first to combine an electron probe with a transmission electron microscope. Figure 3.9 showed the most recent version of his instrument. A "mini" lens (106) (see Section 3.4) focuses the beam onto the specimen and allows room for x-ray spectrometers. If the specimen is partly transparent to electrons (such as particulate matter or a thin substrate or some type of thin film) the following electron lenses form an enlarged shadow image just as in transmission electron microscopes. The advantage is, of course, the better resolution of a transmission microscope compared to even a scanning electron microscope (about 5 Å compared to 100 Å). This allows detailed selection and observation of individual particles which may then be analyzed chemically without removing the specimen from the instrument. Figure 9.14 shows an example of combined microscopy and probe analysis.

A number of manufacturers offer combination instruments* with a variety of electron optics arrangements.

*AEI Scientific Apparatus Ltd., Harlow, Essex, England; CAMECA, Paris, France; TPD-TNO, Delft, Netherlands; Siemens Halske, Munich, Germany.

APPENDIX

1

TABLES OF MASS ATTENUATION COEFFICIENTS FOR $K\alpha$ AND $L\alpha_1$ LINES

The coefficients are given to three significant figures because their values may be that accurate relative to each other but the absolute accuracy is no better than 2% (especially near absorption edges and at long wavelengths).

This set of values was taken from the 1969 tabulation KN-798-69-2(R) by W. J. Veigele, E. Briggs, B. Bracewell, and M. Donaldson, Kaman Nuclear Corp., Colorado Springs, Colorado. Their report contains separate listings for photoelectric absorption, incoherent, and coherent scattering and represents compilation and smoothing of all experimental results published. There is no assumption of a λ^n relationship between absorption coefficient and wavelength, therefore these values should represent an improvement over tabulations which make such an assumption.

The values are shown with the appropriate power of ten, that is, the listing $3.17 + 02$ means 3.17×10^2 or 317 for an iron absorber and $CuK\alpha$ radiation.

MASS ATTENUATION COEFFICIENTS FOR K ALPHA LINES
EMITTER   B   C   N   O   F   NE   NA   MG   AL   SI
WAVELENGTH   67.6   44.7   31.5   23.6   18.3   14.6   11.9   9.8   8.33   7.12

ABSORBER
#	El	B	C	N	O	F	NE	NA	MG	AL	SI
1	H	3.06 +03	9.02 +02	3.08 +02	1.32 +02	6.13 +01	3.11 +01	1.71 +01	1.00 +01	6.06 +00	3.94 +00
2	HE	1.40 +04	4.27 +03	1.48 +03	6.36 +02	2.95 +02	1.48 +02	8.08 +01	4.64 +01	2.70 +01	1.68 +01
3	LI	3.00 +04	1.00 +04	3.73 +03	1.67 +03	8.03 +02	4.14 +02	2.29 +02	1.33 +02	7.84 +01	4.90 +01
4	BE	7.00 +04	2.35 +04	8.82 +03	3.97 +03	1.91 +03	9.91 +02	5.51 +02	3.22 +02	1.89 +02	1.18 +02
5	B	5.42 +04	3.68 +04	1.48 +04	6.99 +03	3.49 +03	1.86 +03	1.05 +03	6.28 +02	3.75 +02	2.38 +02
6	C	6.76 +03	2.37 +04	2.34 +04	1.18 +04	6.23 +03	3.43 +03	1.98 +03	1.19 +03	7.19 +02	4.56 +02
7	N	1.06 +04	3.92 +03	1.64 +04	1.92 +04	9.86 +03	5.37 +03	3.10 +03	1.87 +03	1.13 +03	7.22 +02
8	O	1.65 +04	6.28 +03	2.69 +03	1.34 +04	1.86 +04	7.00 +03	1.05 +03	2.62 +03	1.62 +03	1.06 +03
9	F	2.40 +04	6.93 +03	3.76 +03	1.89 +03	1.43 +03	9.20 +03	4.22 +03	3.29 +03	2.02 +03	1.30 +03
10	NE	3.68 +04	1.33 +04	5.48 +03	2.70 +03	1.43 +03	8.08 +02	5.40 +03	4.56 +03	2.85 +03	1.87 +03
11	NA	5.94 +04	2.08 +04	8.36 +03	4.04 +03	2.10 +03	1.17 +03	6.90 +02	3.29 +03	3.51 +03	2.29 +03
12	MG	9.28 +04	3.13 +04	1.21 +04	5.68 +03	2.87 +03	1.57 +03	9.23 +02	5.56 +02	4.04 +03	2.78 +03
13	AL	1.16 +05	3.85 +04	1.46 +04	6.83 +03	3.45 +03	1.85 +03	1.04 +03	6.61 +02	4.10 +02	3.34 +03
14	SI	1.41 +05	4.79 +04	1.85 +04	8.77 +03	4.45 +03	2.44 +03	1.40 +03	8.88 +02	5.55 +02	3.65 +02
15	P	1.33 +05	5.62 +04	2.21 +04	1.05 +04	5.41 +03	2.99 +03	1.77 +03	1.10 +03	6.93 +02	4.64 +02
16	S	1.41 +05	6.09 +04	2.49 +04	1.22 +04	6.45 +03	3.65 +03	2.21 +03	1.62 +03	8.93 +02	6.93 +02
17	CL	4.14 +04	7.71 +04	3.12 +04	1.50 +04	7.41 +03	4.35 +03	2.60 +03	1.63 +03	1.03 +03	6.93 +02
18	AR	4.97 +04	7.60 +04	3.64 +04	1.73 +04	8.91 +03	4.92 +03	2.92 +03	1.81 +03	1.14 +03	7.62 +02
19	K	6.91 +04	2.34 +04	4.94 +04	2.36 +04	1.51 +04	6.62 +03	3.90 +03	2.42 +03	1.51 +03	1.00 +03
20	CA	7.51 +04	2.86 +04	5.32 +04	2.99 +04	1.72 +04	9.26 +03	5.03 +03	2.98 +03	1.85 +03	1.22 +03
21	SC	1.09 +05	3.62 +04	1.37 +04	3.45 +04	1.88 +04	9.35 +03	5.43 +03	3.32 +03	2.05 +03	1.35 +03
22	TI	1.40 +05	4.56 +04	1.70 +04	3.46 +04	2.07 +04	1.10 +04	6.38 +03	3.86 +03	2.36 +03	1.54 +03
23	V	1.77 +05	5.68 +04	2.08 +04	4.13 +04	2.46 +04	1.31 +04	7.50 +03	4.51 +03	2.74 +03	1.77 +03
24	CR	2.23 +05	6.38 +04	2.56 +04	1.14 +04	2.50 +04	1.55 +04	9.01 +03	5.39 +03	3.26 +03	2.10 +03
25	MN	2.67 +05	1.01 +05	3.65 +04	1.34 +04	1.83 +04	1.83 +04	6.18 +03	6.18 +03	3.72 +03	2.39 +03
26	FE	3.25 +05	1.24 +05	3.94 +04	1.62 +04	7.83 +03	2.17 +04	1.24 +04	7.41 +03	4.46 +03	2.85 +03
27	CO	3.46 +04	2.83 +05	3.46 +04	1.76 +04	8.55 +03	2.02 +04	1.37 +04	7.99 +03	4.94 +03	3.17 +03
28	NI	2.83 +05	3.23 +04	3.46 +04	1.59 +04	4.03 +03	2.34 +03	1.13 +04	8.42 +03	4.90 +03	3.20 +03
29	CU	8.36 +04	1.13 +05	1.41 +05	1.71 +04	1.07 +04	2.16 +04	1.31 +04	8.70 +03	5.17 +03	3.37 +03
30	ZN	3.46 +04	1.20 +05	1.22 +05	1.94 +04	1.21 +04	5.14 +04	8.66 +03	6.56 +03	5.95 +03	3.88 +03
31	GA	3.69 +04	1.47 +05	4.54 +04	2.09 +04	1.26 +04	5.57 +03	1.83 +04	9.70 +03	6.44 +03	4.23 +03
32	GE	4.47 +05	1.48 +05	5.43 +04	2.46 +04	1.31 +04	6.42 +03	3.18 +03	8.56 +03	7.20 +03	4.71 +03
33	AS	4.15 +05	1.62 +05	5.55 +04	2.55 +04	1.51 +04	6.75 +03	3.68 +03	9.64 +03	6.45 +03	4.23 +03
34	SE	8.17 +05	2.40 +05	7.52 +04	3.01 +04	1.86 +04	6.30 +03	3.30 +03	2.31 +03	6.80 +03	5.19 +03
35	BR	4.43 +05	1.27 +05	6.69 +04	3.06 +04	1.86 +04	8.09 +03	4.65 +03	2.82 +03	1.72 +03	5.30 +03
36	KR	8.11 +04	4.33 +04	2.21 +05	1.19 +04	2.04 +04	4.15 +03	2.67 +03	1.79 +03	1.22 +03	4.99 +03
37	RB	4.84 +05	2.05 +05	8.35 +04	3.79 +04	5.80 +04	8.89 +03	5.66 +03	3.41 +03	2.08 +03	1.33 +03
38	SR	5.36 +05	1.97 +05	8.17 +04	4.17 +04	1.45 +04	7.96 +03	4.68 +03	2.89 +03	1.80 +03	1.47 +03
39	Y	3.17 +05	1.07 +05	2.86 +05	2.86 +04	1.54 +04	7.96 +03	4.68 +03	2.89 +03	1.80 +03	1.19 +03
40	ZR	3.49 +05	1.19 +05	6.02 +04	3.08 +04	1.54 +04	9.38 +03	4.89 +03	2.99 +03	1.85 +03	1.21 +03

41	NB	1.09+05	1.35+05	6.71+04	3.45+04	1.72+04	8.31+03	5.41+03	3.30+03	2.04+03	1.34+03
42	MO	9.48+04	1.23+05	6.75+04	3.75+04	1.73+04	8.62+03	5.72+03	3.56+03	2.25+03	1.52+03
43	TC	9.95+04	1.23+05	5.99+04	3.12+04	1.81+04	1.01+04	6.07+03	3.80+03	2.41+03	1.62+03
44	RU	1.02+05	3.40+04	5.29+04	3.37+04	1.03+04	1.08+04	6.51+03	4.09+03	2.60+03	1.75+03
45	RH	1.20+05	4.19+04	5.54+04	2.92+04	2.29+04	1.16+04	6.99+03	4.41+03	2.82+03	1.90+03
46	PD	9.17+04	3.37+04	5.24+04	2.60+04	2.44+04	1.25+04	7.43+03	4.65+03	2.94+03	1.96+03
47	AG	1.22+05	3.38+04	5.23+04	3.26+04	2.22+04	1.16+04	7.45+03	4.56+03	2.95+03	2.01+03
48	CD	1.24+05	4.50+04	1.84+04	3.66+04	2.09+04	1.46+04	8.94+03	5.08+03	3.41+03	2.47+03
49	IN	1.35+05	4.86+04	1.98+04	3.40+04	2.50+04	1.43+04	8.60+03	5.34+03	3.64+03	2.74+03
50	SN	1.96+05	5.49+04	2.75+04	3.04+04	2.11+04	1.63+04	9.60+03	5.67+03	3.90+03	3.06+03
51	SB	1.35+05	7.05+04	2.84+04	1.29+04	2.01+04	1.41+04	1.02+04	6.08+03	4.08+03	3.03+03
52	TE	2.06+05	4.26+04	4.68+04	1.31+04	1.06+04	1.13+04	1.18+04	6.42+03	4.61+03	3.83+03
53	I	1.83+05	1.29+05	4.77+04	9.04+03	1.11+04	9.22+03	9.60+03	7.34+03	4.37+03	3.22+03
54	XE	2.31+05	1.24+05	4.91+04	2.10+04	1.20+04	1.33+04	1.22+04	7.61+03	5.12+03	4.49+03
55	CS	3.25+05	1.17+05	5.22+04	2.17+04	1.25+04	1.63+04	1.13+04	8.06+03	5.94+03	5.02+03
56	BA	3.35+05	1.25+05	5.80+04	2.25+04	1.29+04	2.04+04	1.28+04	8.52+03	6.12+03	5.26+03
57	LA	3.46+05	1.29+05	5.69+04	2.41+04	1.36+04	2.09+04	1.35+04	9.03+03	6.45+03	5.29+03
58	CE	3.54+05	1.22+05	5.55+04	2.49+04	1.45+04	1.91+04	1.40+04	9.29+03	6.83+03	5.49+03
59	PR	3.61+05	1.23+05	5.75+04	2.58+04	1.52+04	2.04+04	1.57+04	9.35+03	6.63+03	5.48+03
60	ND	3.77+05	1.28+05	5.49+04	2.71+04	1.50+04	2.09+04	1.28+04	8.83+03	6.91+03	5.37+03
61	PM	3.72+05	1.34+05	5.75+04	2.94+04	1.70+04	2.91+04	1.41+04	8.29+03	7.24+03	4.68+03
62	SM	3.77+05	1.41+05	6.37+04	2.85+04	1.92+04	2.26+04	1.70+04	9.03+03	6.30+03	4.95+03
63	EU	3.94+05	1.50+05	7.16+04	2.99+04	2.04+04	1.95+04	1.72+04	9.35+03	6.33+03	5.52+03
64	GD	4.41+05	1.72+05	7.42+04	3.34+04	2.12+04	1.05+04	1.50+04	1.01+04	6.76+03	4.75+03
65	TB	6.10+05	1.85+05	7.57+04	3.43+04	2.19+04	1.11+04	1.88+04	7.83+03	7.09+03	5.87+03
66	DY	5.61+05	1.72+05	6.87+04	3.95+04	2.00+04	1.17+04	6.16+03	2.37+03	7.45+03	6.13+03
67	HO	5.50+05	1.56+05	6.54+04	3.78+04	1.76+04	1.21+04	6.52+03	2.50+03	5.27+03	4.75+03
68	ER	5.73+05	1.90+05	6.51+04	3.46+04	1.79+04	1.19+04	6.92+03	2.67+03	2.16+03	1.83+03
69	TM	5.70+05	1.94+05	6.25+04	3.21+04	1.58+04	1.10+04	7.16+03	2.77+03	2.50+03	1.76+03
70	YB	4.99+05	1.72+05	6.18+04	3.06+04	1.59+04	1.09+04	6.61+03	2.66+03	2.67+03	1.74+03
71	LU	4.67+05	1.56+05	5.35+04	3.15+04	1.43+04	1.08+04	6.22+03	2.62+03	2.77+03	1.75+03
72	HF	5.43+05	1.50+05	5.02+04	2.93+04	1.42+04	1.00+04	6.55+03	2.49+03	2.66+03	1.76+03
73	TA	4.67+05	1.63+05	5.18+04	2.59+04	1.64+04	1.08+04	6.46+03	2.60+03	2.59+03	1.74+03
74	W	4.38+05	1.51+05	6.25+04	3.15+04	2.37+04	1.09+04	6.36+03	2.67+03	2.75+03	1.75+03
75	RE	4.20+05	1.55+05	5.35+04	5.69+04	2.67+04	1.55+04	8.54+03	3.91+03	2.89+03	1.92+03
76	OS	2.99+05	2.82+05	1.21+05	5.65+04	2.87+04	1.56+04	9.14+03	4.49+03	2.99+03	2.07+03
77	IR	8.28+05	3.16+05	1.21+05	6.01+04	2.18+04	9.99+03	3.28+03	1.86+03	3.18+03	2.79+03
78	PT	6.02+06	1.76+06	6.03+04	2.56+05	1.18+05	5.99+04	3.28+04	1.86+04	1.21+04	8.41+03

MASS ATTENUATION COEFFICIENTS FOR K ALPHA LINES

EMITTER	P	S	CL	AR	K	CA	SC	TI	V	CR
WAVELENGTH	6.15	5.37	4.72	4.19	3.74	3.35	3.03	2.74	2.50	2.29

ABSORBER

1 H	2.68+00	2.03+00	1.49+00	1.17+00	9.77-01	8.31-01	7.33-01	6.72-01	6.07-01	5.71-01
2 HE	1.04+01	7.64+00	5.04+00	3.50+00	2.58+00	1.90+00	1.45+00	1.18+00	9.06-01	7.59-01
3 LI	3.15+01	2.21+01	1.45+01	9.93+00	7.18+00	5.16+00	3.81+00	3.02+00	2.19+00	1.76+00
4 BE	7.55+01	5.37+01	3.51+01	2.40+01	1.73+01	1.23+01	9.07+00	7.14+00	5.10+00	4.05+00
5 B	1.51+02	1.09+02	7.20+01	4.94+01	3.57+01	2.56+01	1.87+01	1.47+01	1.05+01	8.31+00
6 C	2.94+02	2.08+02	1.37+02	9.34+01	6.71+01	4.77+01	3.46+01	2.70+01	1.90+01	1.49+01
7 N	4.69+02	3.35+02	2.22+02	1.53+02	1.11+02	7.96+01	5.83+01	4.58+01	3.26+01	2.56+01
8 O	7.00+02	5.03+02	3.37+02	2.33+02	1.70+02	1.22+02	8.97+01	7.04+01	5.01+01	3.94+01
9 F	8.56+02	6.16+02	4.12+02	2.86+02	2.08+02	1.50+02	1.11+02	8.74+01	6.27+01	4.95+01
10 NE	1.24+03	9.01+02	6.09+02	4.26+02	3.12+02	2.26+02	1.67+02	1.32+02	9.53+01	7.54+01
11 NA	1.52+03	1.10+03	7.44+02	5.20+02	3.81+02	2.76+02	2.04+02	1.61+02	1.16+02	9.22+01
12 MG	1.93+03	1.42+03	9.93+02	7.11+02	5.29+02	3.90+02	2.92+02	2.32+02	1.69+02	1.34+02
13 AL	2.33+03	1.71+03	1.20+03	8.61+02	6.43+02	4.75+02	3.57+02	2.84+02	2.07+02	1.65+02
14 SI	3.11+03	2.31+03	1.57+03	1.10+03	8.11+02	5.90+02	4.37+02	3.46+02	2.49+02	1.97+02
15 P	3.16+03	2.76+03	1.92+03	1.34+03	9.89+02	7.18+02	5.32+02	4.20+02	3.02+02	2.39+02
16 S	4.21+03	3.51+03	2.38+03	1.67+03	1.23+03	9.01+02	6.69+02	5.30+02	3.83+02	3.04+02
17 CL	4.70+02	3.82+02	2.40+02	1.91+03	1.57+03	1.03+03	7.72+02	6.11+02	4.41+02	3.50+02
18 AR	5.17+02	5.03+02	2.66+02	1.90+02	1.85+03	1.15+03	8.63+02	6.86+02	5.00+02	3.98+02
19 K	6.82+02	5.06+02	3.50+02	2.52+02	2.26+03	1.49+03	1.11+03	8.90+02	6.30+02	4.98+02
20 CA	8.25+02	6.06+02	4.19+02	3.01+02	2.43+02	1.65+03	1.36+03	1.07+03	7.70+02	6.07+02
21 SC	9.00+02	6.61+02	4.55+02	3.25+02	2.48+02	1.80+02	1.34+03	1.07+03	8.20+02	6.55+02
22 TI	1.02+03	7.44+02	5.08+02	3.60+02	2.68+02	1.98+02	1.48+03	1.12+03	8.76+02	6.16+02
23 V	1.17+03	8.48+02	5.75+02	4.05+02	3.00+02	2.21+02	1.66+02	1.32+02	9.64+02	7.57+02
24 CR	1.38+03	9.98+02	6.74+02	4.74+02	3.50+02	2.57+02	1.92+02	1.53+02	1.12+02	8.67+02
25 MN	1.56+03	1.13+03	7.61+02	5.33+02	3.91+02	2.87+02	2.15+02	1.71+02	1.25+02	9.96+01
26 FE	1.87+03	1.34+03	9.06+02	6.34+02	4.67+02	3.41+02	2.54+02	2.02+02	1.48+02	1.18+02
27 CO	2.08+03	1.50+03	1.01+03	7.13+02	5.25+02	3.85+02	2.88+02	2.29+02	1.68+02	1.34+02
28 NI	2.12+03	1.55+03	1.06+03	7.53+02	5.61+02	4.15+02	3.13+02	2.51+02	1.86+02	1.50+02
29 CU	2.24+03	1.63+03	1.12+03	7.96+02	5.92+02	4.38+02	3.31+02	2.65+02	1.96+02	1.58+02
30 ZN	2.58+03	1.87+03	1.28+03	9.11+02	6.78+02	5.01+02	3.78+02	3.03+02	2.24+02	1.80+02
31 GA	2.81+03	2.05+03	1.40+03	9.98+02	7.44+02	5.50+02	4.16+02	3.33+02	2.47+02	1.99+02
32 GE	3.12+03	2.26+03	1.53+03	1.08+03	8.04+02	5.92+02	4.45+02	3.56+02	2.62+02	2.10+02
33 AS	3.38+03	2.46+03	1.68+03	1.19+03	8.87+02	6.56+02	4.94+02	3.96+02	2.93+02	2.36+02
34 SE	3.60+03	2.62+03	1.79+03	1.27+03	9.48+02	7.02+02	5.30+02	4.25+02	3.14+02	2.53+02
35 BR	4.02+03	2.92+03	1.99+03	1.41+03	1.05+03	7.75+02	5.84+02	4.68+02	3.45+02	2.78+02
36 KR	4.29+03	3.12+03	2.13+03	1.50+03	1.12+03	8.27+02	6.23+02	4.99+02	3.67+02	2.96+02
37 RB	3.99+03	3.44+03	2.58+03	1.67+03	1.24+03	9.14+02	6.87+02	5.49+02	4.04+02	3.24+02
38 SR	4.35+03	3.63+03	2.83+03	1.82+03	1.35+03	9.94+02	7.46+02	5.96+02	4.38+02	3.52+02
39 Y	8.06+02	3.30+03	2.83+03	2.00+03	1.48+03	1.08+03	8.17+02	6.52+02	4.79+02	3.84+02
40 ZR	8.16+02	3.44+03	2.97+03	2.11+03	1.57+03	1.15+03	8.69+02	6.95+02	5.11+02	4.11+02

#	El										
41	NB	8.96+02	6.32+02	2.72+02	2.34+03	1.74+03	1.28+03	9.64+02	7.69+02	5.65+02	4.54+02
42	MO	1.02+03	7.58+02	1.95+02	2.14+03	1.84+03	1.38+03	1.06+03	8.26+02	6.05+02	4.85+02
43	TC	1.11+03	8.24+02	5.67+02	4.51+02	1.91+03	1.41+03	1.13+03	8.51+02	6.26+02	5.03+02
44	RU	1.20+03	8.92+02	6.22+02	5.06+02	2.00+03	1.51+03	1.32+03	8.83+02	6.57+02	5.32+02
45	RH	1.31+03	8.76+02	6.97+02	5.56+02	1.24+03	1.74+03	1.37+03	1.05+03	7.78+02	6.24+02
46	PD	1.40+03	9.05+02	6.97+02	6.13+02	4.19+02	1.27+03	1.44+03	1.15+03	8.11+02	6.87+02
47	AG	1.59+03	1.18+03	8.38+02	6.63+02	5.07+02	3.80+02	1.31+03	1.17+03	8.95+02	7.25+02
48	CD	1.70+03	1.27+03	9.04+02	7.18+02	5.41+02	4.13+02	3.14+02	1.19+03	9.28+02	7.53+02
49	IN	1.82+03	1.40+03	9.67+02	7.69+02	5.46+02	4.35+02	3.50+02	7.33+02	9.31+02	7.59+02
50	SN	1.88+03	1.53+03	9.86+02	8.18+02	5.80+02	4.67+02	3.79+02	2.72+02	1.00+03	7.22+02
51	SB	2.07+03	1.61+03	1.06+03	8.75+02	6.12+02	4.95+02	4.29+02	2.97+02	2.05+02	7.58+02
52	TE	2.13+03	1.81+03	1.17+03	8.62+02	6.37+02	5.24+02	4.54+02	3.22+02	2.38+02	8.22+02
53	I	2.51+03	1.61+03	1.29+03	9.77+02	6.73+02	5.35+02	4.86+02	3.50+02	2.61+02	7.58+02
54	XE	2.62+03	1.90+03	1.38+03	1.13+03	7.25+02	6.25+02	5.02+02	3.79+02	2.99+02	5.84+02
55	CS	2.79+03	2.17+03	1.38+03	1.19+03	7.82+02	6.63+02	5.34+02	4.29+02	3.38+02	1.85+02
56	BA	2.98+03	2.32+03	1.59+03	1.33+03	7.44+02	7.44+02	5.65+02	4.54+02	3.62+02	2.27+02
57	LA	3.19+03	2.45+03	1.68+03	1.42+03	8.93+02	5.35+02	6.33+02	4.86+02	3.91+02	2.42+02
58	CE	3.35+03	2.58+03	1.87+03	1.57+03	8.48+02	6.67+02	6.68+02	5.10+02	3.38+02	2.74+02
59	PR	3.71+03	2.71+03	1.99+03	1.62+03	1.06+03	8.32+02	7.37+02	5.64+02	3.62+02	2.93+02
60	ND	3.94+03	2.89+03	2.07+03	1.80+03	1.17+03	8.78+02	7.77+02	5.95+02	4.01+02	3.09+02
61	PM	4.30+03	3.18+03	2.19+03	1.91+03	1.22+03	9.16+02	7.21+02	6.27+02	4.23+02	3.19+02
62	SM	4.08+03	3.22+03	2.26+03	1.98+03	1.35+03	1.01+03	7.96+02	6.62+02	4.46+02	3.25+02
63	EU	4.33+03	3.46+03	2.38+03	2.08+03	1.49+03	1.17+03	8.53+02	6.88+02	4.96+02	3.63+02
64	GD	4.00+03	3.61+03	2.50+03	2.26+03	1.57+03	1.17+03	8.96+02	7.23+02	5.15+02	3.82+02
65	TB	4.21+03	3.24+03	2.65+03	2.35+03	1.64+03	1.27+03	9.33+02	7.82+02	5.40+02	4.18+02
66	DY	4.44+03	3.46+03	2.76+03	2.19+03	1.71+03	1.33+03	9.71+02	8.24+02	5.84+02	4.38+02
67	HO	4.67+03	3.65+03	2.92+03	2.20+03	1.79+03	1.38+03	1.07+03	8.58+02	6.16+02	4.73+02
68	ER	3.93+03	3.74+03	2.82+03	2.26+03	1.87+03	1.43+03	1.06+03	8.91+02	6.46+02	4.55+02
69	TM	4.71+03	3.30+03	2.79+03	2.63+03	1.83+03	1.26+03	1.02+03	8.07+02	6.46+02	5.00+02
70	YB	4.46+03	3.49+03	2.74+03	2.68+03	1.88+03	1.56+03	1.18+03	9.74+02	7.32+02	4.73+02
71	LU	4.65+03	3.30+03	2.96+03	2.83+03	1.82+03	1.43+03	1.20+03	9.94+02	7.47+02	5.26+02
72	HF	2.98+03	2.71+03	2.52+03	7.30+02	1.76+03	1.54+03	1.32+03	1.07+03	8.05+02	5.45+02
73	TA	3.20+03	2.40+03	2.68+03	9.26+02	1.86+03	1.39+03	1.35+03	1.02+03	8.15+02	5.07+02
74	W	1.15+03	7.71+02	2.83+03	7.30+02	1.53+03	1.43+03	1.07+03	1.10+03	7.47+02	5.91+02
75	RE	1.19+03	1.15+03	7.30+02	9.28+02	1.94+03	1.39+03	1.07+03	1.02+03	8.05+02	6.09+02
76	OS	1.20+03	1.20+03	9.58+02	2.56+02	5.16+03	9.22+02	9.02+02	1.02+03	8.15+02	6.76+02
77	IR	1.36+03	1.33+03	1.03+03	9.56+02	1.25+03	9.22+02	1.43+03	1.03+03	8.15+02	6.75+02
78	PT	1.77+03	1.78+03	1.33+03	2.28+02	1.72+03	9.84+02	2.43+03	1.03+03	1.63+03	7.20+02
79	AU	1.88+03	1.78+03	1.33+03	2.51+02	1.72+03	9.84+02	2.43+03	1.81+03	1.63+03	1.26+03

MASS ATTENUATION COEFFICIENTS FOR K ALPHA LINES

EMITTER	MN	FE	CO	NI	CU	ZN	GA	GE	AS	SE
WAVELENGTH	2.10	1.93	1.79	1.65	1.54	1.43	1.34	1.25	1.17	1.10

ABSORBER										
1 H	5.36 -01	5.11 -01	4.89 -01	4.74 -01	4.60 -01	4.50 -01	4.40 -01	4.32 -01	4.25 -01	4.19 -01
2 HE	6.23 -01	5.36 -01	4.60 -01	4.11 -01	3.67 -01	3.37 -01	3.11 -01	2.91 -01	2.76 -01	3.51 -01
3 LI	1.77 +00	1.41 +00	9.01 -01	7.62 -01	6.36 -01	5.08 -01	4.81 -01	4.25 -01	3.86 -01	1.01 -01
4 BE	3.48 +00	2.46 +00	1.93 +00	1.59 +00	1.28 +00	1.08 +00	9.11 -01	7.76 -01	6.85 -01	6.02 +00
5 B	6.28 +00	4.98 +00	3.87 +00	3.15 +00	2.50 +00	2.08 +00	1.83 +00	1.42 +00	1.26 +00	1.69 +00
6 C	1.11 +01	8.77 +00	6.72 +00	5.42 +00	4.24 +00	3.49 +00	2.83 +00	2.32 +00	1.99 +00	2.94 +00
7 N	1.93 +01	1.52 +01	1.17 +01	9.50 +00	7.45 +00	6.13 +00	4.37 +00	4.07 +00	3.47 +00	4.29 +00
8 O	2.95 +01	2.43 +01	1.78 +01	1.15 +01	1.12 +01	9.18 +00	7.30 +00	6.02 +00	6.75 +00	5.66 +00
9 F	3.74 +01	2.96 +01	2.28 +01	1.85 +01	1.45 +01	1.19 +01	9.18 +00	7.92 +00	6.02 +00	8.62 +00
10 NE	5.70 +01	4.41 +01	3.49 +01	2.81 +01	2.22 +01	1.82 +01	1.47 +01	1.20 +01	1.06 +00	1.06 +01
11 NA	6.98 +01	5.54 +01	4.28 +01	3.47 +01	2.73 +01	2.24 +01	1.81 +01	1.48 +01	1.26 +01	1.54 +01
12 MG	1.26 +02	8.14 +01	6.31 +01	5.17 +01	4.01 +01	3.30 +01	2.66 +01	2.17 +01	1.83 +01	1.95 +01
13 AL	1.40 +02	1.18 +02	7.84 +01	7.43 +01	5.82 +01	4.79 +01	3.87 +01	3.16 +01	2.68 +01	2.25 +01
14 SI	1.81 +02	1.83 +02	9.17 +01	9.03 +01	7.08 +01	5.83 +01	4.72 +01	3.85 +01	3.27 +01	2.75 +01
15 P	2.31 +02	2.11 +02	1.42 +02	1.15 +02	1.04 +02	8.59 +01	6.15 +01	4.95 +01	5.82 +01	4.05 +01
16 S	2.65 +02	2.42 +02	1.63 +02	1.32 +02	1.20 +02	8.97 +01	8.98 +01	5.68 +01	5.62 +01	4.72 +01
17 CL	3.73 +02	2.98 +02	1.88 +02	1.53 +02	1.76 +02	1.44 +02	1.16 +02	6.61 +01	8.04 +01	5.55 +01
18 AR	4.35 +02	3.61 +02	2.30 +02	1.86 +02	1.92 +02	1.58 +02	1.28 +02	7.93 +01	8.87 +01	6.74 +01
19 K	3.72 +02	3.32 +02	2.78 +02	2.24 +02	2.07 +02	1.86 +02	1.36 +02	9.49 +01	1.08 +02	7.44 +01
20 CA	4.57 +02	3.92 +02	3.03 +02	2.57 +02	2.74 +02	2.27 +02	1.53 +02	1.26 +02	1.30 +02	9.18 +01
21 SC	4.43 +02	4.28 +02	3.12 +02	2.79 +02	2.23 +02	2.66 +02	1.36 +02	1.45 +02	1.57 +02	1.10 +02
22 TI	4.83 +02	3.94 +02	3.39 +02	3.21 +02	3.05 +02	3.14 +02	2.57 +02	1.70 +02	1.91 +02	1.34 +02
23 V	5.23 +02	4.28 +02	3.39 +02	2.72 +02	3.47 +02	3.14 +02	2.78 +02	2.13 +02	1.91 +02	1.55 +02
24 CR	6.69 +02	5.35 +02	4.67 +02	3.44 +02	4.99 +02	4.43 +02	2.64 +02	2.33 +02	2.01 +02	1.63 +02
25 MN	7.58 +01	6.07 +02	5.73 +02	3.84 +02	5.27 +02	5.56 +02	3.45 +02	2.75 +02	2.54 +02	1.72 +02
26 FE	9.10 +01	7.11 +02	5.73 +02	3.89 +02	5.99 +02	5.75 +02	4.56 +02	3.84 +02	2.37 +02	2.16 +02
27 CO	1.02 +02	8.33 +01	6.56 +02	5.33 +02	6.63 +02	6.54 +02	5.41 +02	3.93 +02	3.74 +02	1.99 +02
28 NI	1.16 +02	9.44 +01	7.49 +01	6.06 +02	7.80 +02	7.06 +02	5.84 +02	4.51 +02	4.15 +02	3.47 +01
29 CU	1.29 +02	1.05 +02	8.49 +01	6.56 +01	8.42 +01	8.54 +02	5.72 +02	5.28 +02	4.83 +02	3.25 +02
30 ZN	1.43 +02	1.25 +02	9.95 +01	8.22 +01	9.15 +01	7.06 +02	7.28 +02	6.06 +02	5.22 +02	4.13 +01
31 GA	1.54 +02	1.31 +02	1.17 +02	8.55 +01	9.72 +01	8.14 +01	7.28 +02	5.28 +02	5.63 +02	4.45 +01
32 GE	1.62 +02	1.49 +02	1.26 +02	1.04 +02	1.05 +02	8.83 +01	6.33 +01	5.51 +01	5.63 +02	4.80 +01
33 AS	1.83 +02	1.59 +02	1.38 +02	1.13 +02	1.14 +02	9.54 +01	6.72 +01	5.88 +01	5.22 +01	5.22 +01
34 SE	1.96 +02	1.74 +02	1.46 +02	1.21 +02	1.24 +02	1.01 +02	7.28 +01	6.06 +01	6.12 +01	5.64 +01
35 BR	2.15 +02	1.85 +02	1.57 +02	1.31 +02	1.31 +02	1.03 +02	7.28 +01	6.54 +01	5.63 +01	5.22 +01
36 KR	2.28 +02	2.02 +02	1.73 +02	1.42 +02	1.46 +02	1.13 +02	7.86 +01	7.12 +01	6.12 +01	5.22 +01
37 RB	2.50 +02	2.19 +02	1.88 +02	1.52 +02	1.57 +02	1.03 +02	7.86 +01	7.12 +01	6.12 +01	5.22 +01
38 SR	2.71 +02	2.39 +02	1.88 +02	1.55 +02	1.24 +02	1.12 +02	7.86 +01	6.54 +01	6.12 +01	5.22 +01
39 Y	2.96 +02	2.56 +02	2.02 +02	1.66 +02	1.34 +02	1.12 +02	9.23 +01	7.63 +01	6.62 +01	5.64 +01
40 ZR	3.17 +02	2.56 +02	2.02 +02	1.66 +02	1.34 +02	1.12 +02	9.23 +01	7.63 +01	6.62 +01	5.64 +01

#	El										
41	NB	3.50+02	2.82+02	2.22+02	1.83+02	1.46+02	1.22+02	1.01+02	8.40+01	7.23+01	6.16+01
42	MO	3.74+02	3.01+02	2.37+02	1.93+02	1.56+02	1.30+02	1.07+02	8.85+01	7.65+01	6.52+01
43	TC	3.88+02	3.13+02	2.47+02	2.03+02	1.63+02	1.36+02	1.12+02	9.35+01	8.04+01	6.86+01
44	RU	4.15+02	3.37+02	2.65+02	2.18+02	1.75+02	1.46+02	1.20+02	1.04+02	8.66+01	7.38+01
45	RH	4.82+02	3.88+02	3.06+02	2.52+02	2.02+02	1.68+02	1.39+02	1.15+02	9.94+01	8.47+01
46	PD	5.06+02	4.09+02	3.23+02	2.66+02	2.14+02	1.79+02	1.48+02	1.23+02	1.06+02	9.04+01
47	AG	5.31+02	4.29+02	3.39+02	2.79+02	2.24+02	1.87+02	1.54+02	1.28+02	1.10+02	9.07+01
48	CD	5.63+02	4.56+02	3.61+02	2.98+02	2.40+02	2.01+02	1.66+02	1.39+02	1.20+02	1.02+02
49	IN	5.83+02	4.72+02	3.73+02	3.08+02	2.47+02	2.07+02	1.71+02	1.42+02	1.23+02	1.05+02
50	SN	5.92+02	4.81+02	3.83+02	3.18+02	2.57+02	2.16+02	1.79+02	1.50+02	1.29+02	1.18+02
51	SB	6.39+02	5.19+02	4.12+02	3.41+02	2.75+02	2.31+02	1.91+02	1.60+02	1.38+02	1.32+02
52	TE	7.25+02	5.93+02	4.53+02	3.74+02	3.01+02	2.52+02	2.08+02	1.73+02	1.49+02	1.50+02
53	I	7.85+02	6.44+02	4.70+02	3.88+02	3.12+02	2.61+02	2.16+02	1.80+02	1.55+02	1.61+02
54	XE	6.71+02	5.72+02	4.53+02	3.88+02	3.38+02	2.83+02	2.34+02	1.95+02	1.68+02	1.72+02
55	CS	7.68+02	6.79+02	5.37+02	4.21+02	3.55+02	2.97+02	2.54+02	2.21+02	1.75+02	1.83+02
56	BA	5.36+02	6.41+02	5.10+02	4.87+02	3.89+02	3.42+02	2.82+02	2.34+02	1.90+02	1.95+02
57	LA	3.16+02	4.58+02	5.70+02	5.11+02	4.39+02	3.66+02	3.23+02	2.68+02	2.15+02	2.04+02
58	CE	3.28+02	4.60+02	5.37+02	5.48+02	4.74+02	3.94+02	3.43+02	2.82+02	2.30+02	2.14+02
59	PR	3.43+02	4.72+02	5.27+02	5.90+02	4.90+02	4.31+02	3.59+02	3.23+02	2.52+02	2.37+02
60	ND	3.55+02	4.86+02	5.97+02	6.00+02	5.17+02	4.58+02	3.95+02	3.43+02	2.68+02	2.50+02
61	PM	3.92+02	5.01+02	6.60+02	6.29+02	4.76+02	4.16+02	3.60+02	3.12+02	2.94+02	2.61+02
62	SM	3.55+02	4.37+02	6.04+02	3.79+02	4.58+02	4.35+02	3.57+02	3.43+02	3.04+02	2.74+02
63	EU	3.92+02	5.36+02	4.36+02	1.35+02	3.26+02	3.09+02	3.95+02	3.76+02	2.95+02	2.89+02
64	GD	4.14+02	5.37+02	1.48+02	1.46+02	1.32+02	3.16+02	3.00+02	3.30+02	3.18+02	2.94+02
65	TB	4.25+02	5.39+02	1.55+02	1.63+02	1.45+02	1.27+02	2.37+02	2.37+02	2.13+02	3.00+02
66	DY	4.68+02	5.69+02	1.76+02	1.72+02	1.51+02	1.31+02	1.14+02	1.88+02	2.09+02	2.36+02
67	HO	4.79+02	6.00+02	1.96+02	1.78+02	1.56+02	1.36+02	1.21+02	1.08+02	2.13+02	1.83+02
68	ER	5.16+02	3.16+02	2.07+02	1.86+02	1.63+02	1.46+02	1.28+02	1.14+02	9.45+01	1.96+02
69	TM	5.33+02	3.28+02	2.24+02	1.92+02	1.73+02	1.54+02	1.31+02	1.17+02	1.02+02	8.76+01
70	YB	5.37+02	3.43+02	2.41+02	2.00+02	1.85+02	1.57+02	1.35+02	1.24+02	1.07+02	9.44+01
71	LU	6.44+02	3.55+02	2.56+02	2.13+02	1.91+02	1.61+02	1.37+02	1.27+02	1.10+02	9.56+01
72	HF	1.06+03	3.92+02	2.73+02	2.27+02	1.90+02	1.78+02	1.47+02	1.38+02	1.24+02	1.07+02
73	TA	4.23+02	2.81+02	2.30+02	2.08+02	1.76+02	1.51+02	1.43+02	1.33+02	1.15+02
74	W	4.37+02	3.08+02	2.57+02	2.14+02	1.95+02	1.63+02	1.52+02	1.40+02	1.40+02
75	RE	4.43+02	3.15+02	2.62+02	2.31+02	2.13+02	1.79+02	1.86+02	1.63+02	1.82+02
76	OS	5.38+02	3.40+02	2.83+02	2.49+02	2.53+02	2.18+02	2.54+02	2.15+02		
77	IR	5.45+02	3.60+02	2.93+02	3.04+02	3.11+02					
78	PT	6.44+02	3.40+03	3.68+02	4.65+02						
92	TH		3.70+02	5.92+02							
94	PU		4.45+02								

MASS ATTENUATION COEFFICIENTS FOR K ALPHA LINES

EMITTER	BR	KR	RB	SR	Y	ZR	NB	MO
WAVELENGTH	1.04	.981	.926	.876	.830	.787	.747	.710

ABSORBER
1 H	4.12 -01	4.08 -01	4.05 -01	4.01 -01	3.97 -01	3.94 -01	3.91 -01	3.88 -01
2 HE	2.49 -01	2.42 -01	2.35 -01	2.28 -01	2.22 -01	2.17 -01	2.13 -01	2.09 -01
3 LI	3.15 -01	2.96 -01	2.79 -01	2.62 -01	2.44 -01	2.33 -01	2.23 -01	2.14 -01
4 BE	5.17 -01	4.72 -01	4.32 -01	3.91 -01	3.48 -01	3.24 -01	3.01 -01	2.83 -01
5 B	8.89 -01	7.95 -01	7.12 -01	6.27 -01	5.39 -01	4.89 -01	4.43 -01	4.06 -01
6 C	1.38 +00	1.22 +00	1.08 +00	9.38 -01	7.88 -01	7.05 -01	6.27 -01	5.66 -01
7 N	2.38 +00	2.09 +00	1.83 +00	1.56 +00	1.29 +00	1.14 +00	9.98 -01	8.84 -01
8 O	3.44 +00	3.00 +00	2.61 +00	2.21 +00	1.79 +00	1.57 +00	1.36 +00	1.19 +00
9 F	4.58 +00	4.00 +00	3.48 +00	2.95 +00	2.40 +00	2.10 +00	1.81 +00	1.58 +00
10 NE	5.93 +00	6.04 +00	5.24 +00	4.42 +00	3.58 +00	3.11 +00	2.67 +00	2.33 +00
11 NA	8.57 +00	7.46 +00	6.48 +00	5.46 +00	4.42 +00	3.84 +00	3.30 +00	2.87 +00
12 MG	1.24 +01	1.07 +01	9.31 +00	7.82 +00	6.27 +00	5.43 +00	4.63 +00	4.00 +00
13 AL	1.57 +01	1.36 +01	1.18 +01	9.92 +00	7.97 +00	6.90 +00	5.89 +00	5.08 +00
14 SI	1.80 +01	1.57 +01	1.36 +01	1.14 +01	9.18 +00	7.95 +00	6.79 +00	5.86 +00
15 P	2.21 +01	1.92 +01	1.66 +01	1.39 +01	1.12 +01	9.75 +00	8.33 +00	7.19 +00
16 S	2.84 +01	2.46 +01	2.13 +01	1.79 +01	1.44 +01	1.25 +01	1.06 +01	9.22 +00
17 CL	3.25 +01	2.83 +01	2.45 +01	2.06 +01	1.65 +01	1.43 +01	1.22 +01	1.05 +01
18 AR	3.80 +01	3.30 +01	2.86 +01	2.40 +01	1.93 +01	1.67 +01	1.42 +01	1.23 +01
19 K	4.54 +01	3.94 +01	3.41 +01	2.87 +01	2.30 +01	2.00 +01	1.70 +01	1.47 +01
20 CA	5.40 +01	4.68 +01	4.05 +01	3.40 +01	2.72 +01	2.35 +01	2.01 +01	1.73 +01
21 SC	5.97 +01	5.18 +01	4.49 +01	3.77 +01	3.02 +01	2.61 +01	2.23 +01	1.92 +01
22 TI	6.96 +01	6.08 +01	5.29 +01	4.48 +01	3.64 +01	3.16 +01	2.71 +01	2.35 +01
23 V	7.47 +01	6.52 +01	5.67 +01	4.80 +01	3.90 +01	3.39 +01	2.90 +01	2.51 +01
24 CR	8.94 +01	7.80 +01	6.77 +01	5.71 +01	4.62 +01	4.01 +01	3.43 +01	2.96 +01
25 MN	9.22 +01	8.65 +01	7.51 +01	6.33 +01	5.12 +01	4.44 +01	3.80 +01	3.28 +01
26 FE	1.10 +02	9.64 +01	8.41 +01	7.08 +01	5.84 +01	5.09 +01	4.38 +01	3.80 +01
27 CO	1.27 +02	1.11 +02	9.66 +01	8.17 +01	6.64 +01	5.77 +01	4.94 +01	4.27 +01
28 NI	1.34 +02	1.18 +02	1.03 +02	8.81 +01	7.24 +01	6.33 +01	5.46 +01	4.75 +01
29 CU	1.42 +02	1.25 +02	1.09 +02	9.33 +01	7.67 +01	6.71 +01	5.79 +01	5.04 +01
30 ZN	1.65 +02	1.44 +02	1.26 +02	1.07 +02	8.74 +01	7.62 +01	6.55 +01	5.68 +01
31 GA	1.77 +02	1.55 +02	1.34 +02	1.13 +02	9.16 +01	7.95 +01	6.80 +01	5.88 +01
32 GE	1.69 +02	1.49 +02	1.31 +02	1.12 +02	9.21 +01	8.05 +01	6.95 +01	6.06 +01
33 AS	2.12 +02	1.86 +02	1.61 +02	1.36 +02	1.10 +02	9.56 +01	8.17 +01	7.06 +01
34 SE	2.96 +02	2.52 +02	2.15 +02	1.80 +02	1.44 +02	1.24 +02	1.05 +02	9.04 +01
35 BR	3.18 +01	2.76 +01	2.34 +01	1.51 +02	1.16 +02	1.01 +02	8.75 +01	7.59 +01
36 KR	3.40 +01	2.88 +01	2.52 +01	2.15 +01	1.27 +02	1.11 +02	9.57 +01	8.29 +01
37 RB	3.66 +01	3.14 +01	2.66 +01	2.32 +01	1.98 +01	1.12 +02	9.70 +01	8.43 +01
38 SR	3.94 +01	3.43 +01	2.95 +01	2.46 +01	2.13 +01	1.85 +01	1.14 +02	9.89 +01
39 Y	4.28 +01	3.76 +01	3.29 +01	2.81 +01	2.31 +01	2.04 +01	1.76 +01	1.07 +02
40 ZR	4.64 +01	4.08 +01	3.58 +01	3.06 +01	2.52 +01	2.18 +01	1.91 +01	1.66 +01

#	El								
41	NB	5.06+01	4.44+01	3.89+01	3.32+01	2.74+01	2.39+01	2.06+01	1.81+01
42	MO	5.34+01	4.69+01	4.11+01	3.50+01	2.88+01	2.53+01	2.19+01	1.88+01
43	TC	5.63+01	4.95+01	4.34+01	3.70+01	3.05+01	2.68+01	2.32+01	2.01+01
44	RU	6.06+01	5.33+01	4.67+01	3.99+01	3.29+01	2.89+01	2.51+01	2.19+01
45	RH	6.42+01	5.66+01	4.98+01	4.27+01	3.55+01	3.12+01	2.72+01	2.39+01
46	PD	6.95+01	6.11+01	5.35+01	4.56+01	3.76+01	3.30+01	2.85+01	2.50+01
47	AG	7.46+01	6.56+01	5.76+01	4.92+01	4.06+01	3.57+01	3.09+01	2.71+01
48	CD	7.75+01	6.81+01	5.97+01	5.10+01	4.20+01	3.69+01	3.20+01	2.80+01
49	IN	8.46+01	7.45+01	6.56+01	5.70+01	4.70+01	4.13+01	3.58+01	3.14+01
50	SN	8.63+01	7.60+01	6.66+01	5.70+01	4.72+01	4.13+01	3.58+01	3.14+01
51	SB	9.20+01	8.11+01	7.13+01	6.12+01	5.08+01	4.48+01	3.90+01	3.42+01
52	TE	9.76+01	8.26+01	7.56+01	6.48+01	5.37+01	4.72+01	4.10+01	3.60+01
53	I	1.05+02	9.26+01	8.12+01	6.94+01	5.73+01	5.04+01	4.37+01	3.83+01
54	XE	1.18+02	9.62+01	8.10+01	6.94+01	5.96+01	5.24+01	4.55+01	3.98+01
55	CS	1.23+02	1.08+02	9.49+01	7.78+01	6.42+01	5.86+01	4.89+01	4.28+01
56	BA	1.18+02	1.15+02	1.08+02	8.11+01	6.68+01	5.67+01	5.08+01	4.44+01
57	LA	1.41+02	1.23+02	1.08+02	8.62+01	7.07+01	6.13+01	5.34+01	4.66+01
58	CE	1.50+02	1.32+02	1.15+02	9.25+01	7.61+01	6.67+01	5.77+01	5.05+01
59	PR	1.59+02	1.39+02	1.22+02	9.88+01	8.12+01	7.12+01	6.16+01	5.38+01
60	ND	1.67+02	1.46+02	1.28+02	1.03+02	8.50+01	7.43+01	6.45+01	5.58+01
61	PM	1.78+02	1.53+02	1.34+02	1.09+02	8.96+01	7.84+01	6.78+01	5.91+01
62	SM	1.86+02	1.68+02	1.47+02	1.21+02	9.88+01	8.20+01	7.08+01	6.18+01
63	EU	1.92+02	1.78+02	1.56+02	1.25+02	1.03+02	8.68+01	7.49+01	6.53+01
64	GD	2.13+02	1.86+02	1.63+02	1.32+02	1.09+02	8.94+01	7.71+01	6.72+01
65	TB	2.24+02	1.96+02	1.71+02	1.38+02	1.13+02	9.38+01	8.17+01	7.12+01
66	DY	2.37+02	2.08+02	1.80+02	1.45+02	1.19+02	9.47+01	8.52+01	7.42+01
67	HO	2.45+02	2.15+02	1.81+02	1.52+02	1.26+02	1.04+02	8.99+01	7.83+01
68	ER	2.37+02	2.22+02	1.94+02	1.60+02	1.31+02	1.09+02	9.38+01	8.17+01
69	TM	2.52+02	2.33+02	2.04+02	1.65+02	1.36+02	1.15+02	9.92+01	8.65+01
70	YB	2.64+02	2.19+02	1.92+02	1.73+02	1.42+02	1.19+02	9.58+01	8.65+01
71	LU	2.47+02	2.45+02	2.16+02	1.64+02	1.35+02	1.24+02	1.02+02	9.04+01
72	HF	1.61+02	2.07+02	1.87+02	1.85+02	1.59+02	1.29+02	1.03+02	9.37+01
73	TA	1.65+02	1.56+02	1.84+02	1.95+02	1.59+02	1.36+02	1.12+02	1.03+02
74	W	1.72+02	1.61+02	1.39+02	1.96+02	1.68+02	1.40+02	1.18+02	1.10+02
75	RE	1.87+02	1.83+02	1.38+02	1.17+02	1.79+02	1.48+02	1.26+02	1.12+02
76	OS	7.74+01	7.82+01	1.57+02	1.23+02	1.54+02	1.58+02	1.28+02	1.13+02
77	IR	7.97+01	7.82+01	1.42+02	1.38+02	1.48+02	1.61+02	1.37+02	1.20+02
78	PT	8.64+01	8.64+01	1.36+02	6.58+01	1.16+02	1.44+02	1.41+02	1.23+02
79	AU	8.99+01	7.05+01	1.57+02	8.23+01	1.17+02	1.04+02	1.50+02	1.32+02
80	HG	9.74+01	6.83+01	6.70+01	8.23+01	6.97+01	6.04+01	9.07+01	1.30+02
81	TL	8.99+01	7.82+01	7.63+01	9.40+01	7.61+01	6.21+01	5.47+01	1.16+02
82	PB	1.05+02	8.64+01	9.32+01	8.23+01	1.17+02	6.61+01	5.66+01	9.81+01
83	BI	1.19+02	1.05+02	9.45+01	8.23+01	6.97+01	6.21+01	5.47+01	4.85+01
84	PO	1.47+02	1.28+02	1.11+02	9.40+01	7.61+01	6.61+01	5.66+01	-

MASS ATTENUATION COEFFICIENTS FOR L ALPHA 1 LINES

EMITTER	CA	SC	TI	V	CR	MN	FE	CO	NI	CU
WAVELENGTH	36.3	31.3	27.4	24.2	21.6	19.4	17.5	15.9	14.5	13.3

ABSORBER

1 H	4.70 +02	2.99 +02	2.14 +02	1.40 +02	1.02 +02	7.37 +01	5.31 +01	4.05 +01	3.08 +01	2.35 +01
2 HE	2.25 +03	1.44 +03	1.03 +03	6.77 +02	4.93 +02	3.55 +02	2.55 +02	1.94 +02	1.46 +02	1.11 +02
3 LI	5.53 +03	3.64 +03	2.64 +03	1.77 +03	1.31 +03	9.60 +02	6.99 +02	5.37 +02	4.10 +02	3.15 +02
4 BE	1.30 +04	8.60 +03	6.25 +03	4.22 +03	3.12 +03	2.28 +03	1.67 +03	1.28 +03	9.82 +02	7.55 +02
5 B	2.13 +04	1.44 +04	1.07 +04	7.40 +03	5.55 +03	4.13 +03	3.06 +03	2.38 +03	1.84 +03	1.43 +03
6 C	3.22 +04	2.29 +04	1.74 +04	1.25 +04	9.60 +03	7.30 +03	5.52 +03	4.34 +03	3.40 +03	2.67 +03
7 N	2.31 +03	1.60 +03	2.86 +03	2.02 +03	1.54 +03	1.16 +03	8.71 +03	6.82 +03	5.32 +03	4.17 +03
8 O	3.76 +03	2.63 +03	2.00 +03	1.43 +03	9.44 +03	7.16 +03	1.08 +04	8.71 +03	6.94 +03	5.55 +03
9 F	5.29 +03	3.68 +03	2.78 +03	1.99 +03	1.78 +03	1.40 +03	1.47 +04	1.16 +04	9.12 +03	7.20 +03
10 NE	7.79 +03	5.36 +03	4.03 +03	2.85 +03	1.53 +03	1.17 +03	1.27 +03	1.01 +04	8.00 +03	9.62 +03
11 NA	1.20 +04	8.17 +03	6.09 +03	4.27 +03	2.19 +03	1.67 +03	1.86 +03	1.47 +03	1.16 +03	9.25 +02
12 MG	1.76 +04	1.18 +04	8.71 +03	6.01 +03	3.25 +03	2.45 +03	2.53 +03	1.99 +03	1.56 +03	1.22 +03
13 AL	2.15 +04	1.43 +04	1.05 +04	7.22 +03	4.53 +03	3.39 +03	3.01 +03	2.35 +03	1.83 +03	1.44 +03
14 SI	2.70 +04	1.81 +04	1.34 +04	9.27 +03	5.42 +03	5.24 +03	3.93 +03	3.08 +03	2.42 +03	1.91 +03
15 P	3.19 +04	2.16 +04	1.60 +04	1.11 +04	7.00 +03	6.35 +03	5.78 +03	3.77 +03	2.96 +03	2.35 +03
16 S	3.54 +04	2.43 +04	1.82 +04	1.29 +04	8.44 +03	2.98 +03	5.73 +03	4.56 +03	3.62 +03	2.89 +03
17 CL	4.49 +04	3.05 +04	2.27 +04	1.59 +04	9.88 +03	2.64 +03	2.92 +03	5.47 +03	4.32 +03	3.43 +03
18 AR	5.21 +04	3.55 +04	2.63 +04	1.83 +04	1.21 +04	7.54 +03	7.87 +03	6.21 +03	4.88 +03	3.86 +03
19 K	6.01 +04	4.83 +04	3.59 +04	2.49 +04	1.39 +04	9.15 +03	1.06 +04	8.36 +03	6.56 +03	5.18 +03
20 CA	1.70 +04	5.14 +04	4.39 +04	3.17 +04	1.88 +04	1.41 +04	1.33 +04	1.04 +04	8.19 +03	6.45 +03
21 SC	2.01 +04	1.34 +04	4.40 +04	3.69 +04	2.38 +04	1.78 +04	1.52 +04	1.18 +04	9.27 +03	7.27 +03
22 TI	2.51 +04	1.66 +04	1.14 +04	3.69 +04	2.74 +04	2.04 +04	1.82 +04	1.41 +04	1.09 +04	8.58 +03
23 V	3.09 +04	2.03 +04	1.47 +04	9.84 +03	3.22 +04	2.46 +04	2.17 +04	1.68 +04	1.30 +04	1.01 +04
24 CR	3.82 +04	2.50 +04	1.81 +04	1.21 +04	3.30 +04	2.89 +04	2.64 +04	2.04 +04	1.57 +04	1.23 +04
25 MN	4.52 +04	2.94 +04	2.13 +04	1.43 +04	8.76 +03	2.98 +04	2.55 +04	2.32 +04	1.82 +04	1.41 +04
26 FE	5.47 +04	3.56 +04	2.57 +04	1.73 +04	1.05 +04	7.54 +03	6.81 +03	2.35 +04	2.15 +04	1.69 +04
27 CO	5.89 +04	3.84 +04	2.78 +04	1.87 +04	1.38 +04	1.01 +04	7.46 +03	5.68 +03	2.00 +04	1.84 +04
28 NI	5.10 +04	3.37 +04	2.47 +04	1.68 +04	1.26 +04	9.34 +03	6.92 +03	5.39 +03	4.12 +03	1.47 +04
29 CU	1.96 +04	1.38 +04	1.05 +04	7.63 +03	5.99 +03	4.65 +03	3.52 +03	2.93 +03	2.00 +03	1.90 +03
30 ZN	6.23 +04	1.66 +04	1.14 +04	2.05 +04	1.53 +04	1.13 +04	8.43 +03	6.56 +03	5.10 +03	3.98 +03
31 GA	6.68 +04	4.12 +04	3.01 +04	2.21 +04	1.65 +04	1.22 +04	9.11 +03	7.10 +03	5.52 +03	4.31 +03
32 GE	3.82 +04	4.42 +04	3.24 +04	2.61 +04	1.94 +04	1.43 +04	1.05 +04	8.22 +03	6.36 +03	4.96 +03
33 AS	8.19 +04	5.29 +04	3.86 +04	2.70 +04	2.01 +04	1.49 +04	1.10 +04	8.62 +03	6.69 +03	5.23 +03
34 SE	1.18 +04	5.42 +04	3.96 +04	3.21 +04	2.29 +04	1.32 +04	1.12 +04	8.40 +03	6.23 +03	4.66 +03
35 BR	9.88 +04	6.53 +04	4.77 +04	3.24 +04	2.42 +04	1.60 +04	1.10 +04	1.03 +04	8.01 +03	6.26 +03
36 KR	3.01 +04	2.17 +04	1.68 +04	1.24 +04	9.87 +03	1.79 +04	1.32 +04	5.03 +03	6.11 +03	6.38 +03
37 RB	1.22 +05	8.14 +04	5.93 +04	4.42 +04	3.29 +04	2.21 +04	1.63 +04	1.26 +04	9.80 +03	7.63 +03
38 SR	1.24 +04	8.95 +04	6.53 +04	4.42 +04	3.29 +04	2.42 +04	1.63 +04	1.39 +04	1.07 +04	8.37 +03
39 Y	7.97 +04	5.88 +04	4.37 +04	3.02 +04	2.28 +04	1.71 +04	1.28 +04	1.00 +04	7.89 +03	6.22 +03
40 ZR	7.87 +04	5.87 +04	4.57 +04	3.26 +04	2.45 +04	1.82 +04	1.36 +04	1.06 +04	8.30 +03	6.52 +03

#	El	C1	C2	C3	C4	C5	C6	C7	C8	C9	C10
41	NB	7.23+03	9.23+03	1.18+04	1.51+04	2.03+04	2.74+04	3.65+04	4.66+04	6.56+04	7.51+04
42	MO	7.56+03	9.54+03	1.21+04	1.53+04	2.03+04	2.69+04	3.54+04	4.59+04	5.64+04	6.88+04
43	TC	7.99+03	1.00+04	1.27+04	1.61+04	2.12+04	2.75+04	3.33+04	4.56+04	5.87+04	7.17+04
44	RU	8.56+03	1.07+04	1.36+04	1.72+04	2.26+04	2.68+04	3.56+04	3.69+04	5.17+04	7.58+04
45	RH	9.17+03	1.15+04	1.45+04	1.82+04	2.85+04	3.09+04	3.13+04	3.53+04	5.42+04	7.88+04
46	PD	9.83+03	1.24+04	1.57+04	1.99+04	2.37+04	2.48+04	2.75+04	3.88+04	6.10+04	8.85+04
47	AG	9.28+03	1.39+04	1.72+04	1.64+04	2.34+04	2.63+04	3.44+04	3.53+04	5.13+04	8.49+04
48	CD	1.10+04	1.45+04	1.67+04	2.00+04	2.06+04	2.74+04	3.58+04	4.53+04	1.80+04	1.98+04
49	IN	1.17+04	1.52+04	1.74+04	2.10+04	2.25+04	3.17+04	3.86+04	4.81+04	1.94+04	2.54+04
50	SN	1.25+04	1.62+04	1.78+04	1.85+04	2.44+04	2.53+04	1.38+04	1.43+04	2.15+04	2.62+04
51	SB	1.21+04	1.12+04	1.39+04	2.27+04	2.94+04	7.69+03	1.41+04	1.61+04	2.69+04	2.83+04
52	TE	1.42+04	1.98+04	2.55+04	1.72+04	2.95+04	7.65+03	9.90+03	1.99+04	1.80+04	3.16+04
53	I	1.32+04	2.02+04	6.88+03	1.86+04	5.90+03	1.66+04	2.22+04	1.37+04	4.56+04	3.05+04
54	XE	1.53+04	2.07+04	7.26+03	8.00+03	5.21+03	1.66+04	2.23+04	2.03+04	4.54+04	2.56+04
55	CS	9.00+03	5.85+03	7.59+03	9.35+03	1.21+04	1.71+04	2.30+04	3.31+04	4.65+04	6.93+04
56	BA	1.57+04	6.69+03	7.98+03	9.75+03	1.31+04	1.78+04	2.38+04	3.39+04	4.79+04	7.07+04
57	LA	1.64+04	6.42+03	8.25+03	1.02+04	1.38+04	1.86+04	2.48+04	3.50+04	4.97+04	7.25+04
58	CE	1.72+04	7.48+03	8.59+03	1.05+04	1.42+04	1.91+04	2.64+04	3.64+04	5.09+04	7.49+04
59	PR	4.80+03	6.96+03	8.92+03	1.11+04	1.47+04	1.98+04	2.73+04	3.73+04	5.25+04	7.60+04
60	ND	5.01+03	7.34+03	9.40+03	1.18+04	1.53+04	2.05+04	2.87+04	3.98+04	5.41+04	7.76+04
61	PM	5.25+03	8.00+03	9.56+03	1.31+04	1.61+04	2.15+04	2.89+04	4.17+04	5.66+04	7.32+04
62	SM	5.46+03	7.91+03	1.02+04	1.33+04	1.63+04	2.18+04	3.02+04	4.52+04	5.67+04	7.73+04
63	EU	5.77+03	8.30+03	1.05+04	1.41+04	1.74+04	2.34+04	3.11+04	4.27+04	5.36+04	7.74+04
64	GD	5.89+03	8.73+03	1.11+04	1.47+04	1.79+04	2.28+04	3.16+04	4.54+04	5.88+04	8.30+04
65	TB	6.29+03	9.27+03	1.18+04	1.48+04	1.88+04	2.39+04	3.25+04	4.20+04	6.50+04	7.86+04
66	DY	6.24+03	1.01+04	1.31+04	1.53+04	1.89+04	2.51+04	3.32+04	4.52+04	5.55+04	7.49+04
67	HO	6.89+03	1.10+04	1.33+04	1.42+04	1.95+04	2.67+04	3.16+04	4.27+04	5.36+04	8.49+04
68	ER	6.55+03	1.20+04	1.41+04	1.37+04	1.80+04	3.01+04	3.54+04	4.54+04	5.88+04	8.49+04
69	TM	7.32+03	1.16+04	1.47+04	1.28+04	1.73+04	3.12+04	3.97+04	4.74+04	6.98+04	1.06+05
70	YB	7.93+03	1.12+04	1.48+04	1.25+04	1.48+04	3.16+04	3.85+04	5.21+04	5.05+04	9.69+04
71	LU	8.20+03	1.09+04	1.42+04	1.12+04	1.58+04	3.25+04	4.03+04	5.39+04	7.24+04	1.03+05
72	HF	8.69+03	1.02+04	1.37+04	1.22+04	2.06+04	3.04+04	3.74+04	5.12+04	7.22+04	1.08+05
73	TA	9.05+03	1.05+04	1.38+04	1.28+04	1.75+04	3.12+04	4.06+04	5.29+04	6.67+04	1.07+05
74	W	9.20+03	1.05+04	1.28+04	1.22+04	1.85+04	2.82+04	3.85+04	5.12+04	6.95+04	1.09+05
75	RE	9.50+03	9.10+03	1.25+04	1.03+04	1.66+04	2.53+04	3.43+04	5.29+04	6.19+04	9.86+04
76	OS	8.87+03	9.15+03	1.12+04	1.18+04	1.94+04	2.36+04	3.33+04	4.95+04	5.90+04	9.06+04
77	IR	8.70+03	9.47+03	1.22+04	1.49+04	1.43+04	2.44+04	3.09+04	4.63+04	6.10+04	8.66+04
78	PT	8.68+03	—	1.98+04	2.33+04	2.15+04	2.15+04	3.20+04	4.41+04	6.19+04	8.95+04
79	AU	8.18+03	—	—	2.83+04	2.13+04	2.55+04	2.77+04	4.56+04	6.07+04	7.25+04
80	HG	8.50+03	—	—	—	2.06+04	2.42+04	3.32+04	3.85+04	4.92+04	6.98+04
81	TL	8.11+03	1.54+04	—	—	1.75+04	2.12+04	3.09+04	3.67+04	6.21+04	6.00+04
82	PB	8.64+03	—	—	—	1.85+04	2.55+04	3.20+04	4.74+04	5.07+04	8.36+04
83	BI	8.27+03	—	1.83+04	—	1.66+04	4.15+04	3.49+04	7.93+04	1.07+05	1.59+05
86	RN	7.36+03	—	—	—	1.94+04	4.53+04	6.01+04	3.67+04	1.07+05	1.77+05
90	TH	1.13+04	1.54+04	1.83+04	2.33+04	3.11+04	4.52+04	3.32+04	7.93+04	5.84+04	9.19+04
92	U	1.21+04	1.54+04	1.98+04	2.33+04	3.38+04	4.52+04	5.49+04	8.73+04	—	—
94	PU	4.52+04	5.93+04	7.82+04	1.02+05	1.43+05	1.98+05	2.72+05	4.16+05	—	—

MASS ATTENUATION COEFFICIENTS FOR L ALPHA 1 LINES

EMITTER	ZN	GA	GE	AS	SE	BR	KR	RB	SR	Y
WAVELENGTH	12.2	11.2	10.4	9.67	8.98	8.37	7.81	7.31	6.86	6.44

ABSORBER

1 H	1.83 +01	1.49 +01	1.14 +01	9.60 +00	7.92 +00	6.17 +00	5.10 +00	4.18 +00	3.63 +00	3.07 +00
2 HE	8.63 +01	7.01 +01	5.32 +01	4.41 +01	3.60 +01	2.75 +01	2.24 +01	1.79 +01	1.53 +01	1.26 +01
3 LI	2.45 +02	1.99 +02	1.52 +02	1.27 +02	1.04 +02	7.99 +01	6.51 +01	5.24 +01	4.46 +01	3.68 +01
4 BE	5.88 +02	4.80 +02	3.67 +02	3.05 +02	2.50 +02	1.93 +02	1.57 +02	1.26 +02	1.08 +02	8.93 +01
5 B	2.11 +03	9.24 +02	7.14 +02	5.97 +02	4.92 +02	3.82 +02	3.13 +02	2.53 +02	2.16 +02	1.80 +02
6 C	2.12 +03	1.74 +03	1.35 +03	1.13 +03	9.39 +02	7.31 +02	6.01 +02	4.87 +02	4.16 +02	3.45 +02
7 N	3.30 +03	2.72 +03	2.12 +03	1.78 +03	1.47 +03	1.15 +03	9.47 +02	7.70 +02	6.59 +02	5.49 +02
8 O	4.47 +03	3.73 +03	2.95 +03	2.50 +03	2.08 +03	1.65 +03	1.37 +03	1.12 +03	9.71 +02	8.13 +02
9 F	5.73 +03	4.75 +03	3.72 +03	3.14 +03	2.61 +03	2.05 +03	1.69 +03	1.39 +03	1.19 +03	9.98 +02
10 NE	7.75 +03	6.47 +03	5.14 +03	4.36 +03	3.64 +03	2.89 +03	2.41 +03	1.98 +03	1.71 +03	1.44 +03
11 NA	7.40 +03	7.85 +03	6.36 +03	5.40 +03	4.51 +03	3.57 +03	2.97 +03	2.44 +03	2.10 +03	1.76 +03
12 MG	9.79 +02	8.14 +03	6.41 +03	5.22 +03	4.87 +03	4.04 +03	3.49 +03	2.94 +03	2.57 +03	2.19 +03
13 AL	1.14 +03	9.51 +02	7.46 +03	6.30 +03	5.26 +03	4.16 +03	4.15 +03	3.53 +03	3.08 +03	2.64 +03
14 SI	1.52 +03	1.26 +03	1.00 +03	8.47 +03	7.09 +02	5.64 +03	4.71 +03	3.90 +03	3.31 +03	3.62 +03
15 P	1.88 +03	1.56 +03	1.26 +03	1.05 +03	8.84 +02	7.06 +02	5.92 +02	4.91 +02	5.62 +03	4.80 +03
16 S	2.34 +03	1.96 +03	1.57 +03	1.34 +03	1.13 +03	9.13 +02	7.70 +02	6.44 +02	6.38 +03	5.43 +03
17 CL	2.75 +03	2.30 +03	1.83 +03	1.55 +03	1.31 +03	1.05 +03	8.82 +02	7.34 +02	7.00 +03	5.94 +03
18 AR	3.09 +03	2.58 +03	2.04 +03	1.73 +03	1.45 +03	1.16 +03	9.72 +02	8.07 +02	1.12 +03	7.84 +03
19 K	4.14 +03	3.45 +03	2.72 +03	2.31 +03	1.93 +03	1.54 +03	1.28 +03	1.06 +03	1.12 +03	9.51 +02
20 CA	5.13 +03	3.36 +03	3.75 +03	2.84 +03	2.37 +03	1.88 +03	1.57 +03	1.30 +03	1.23 +03	1.04 +03
21 SC	5.77 +03	4.78 +03	3.31 +03	3.17 +03	2.64 +03	2.09 +03	1.74 +03	1.43 +03	1.41 +03	1.18 +03
22 TI	6.78 +03	5.60 +03	4.37 +03	3.68 +03	3.05 +03	2.40 +03	1.99 +03	1.63 +03	1.62 +03	1.36 +03
23 V	7.98 +03	6.58 +03	5.11 +03	4.29 +03	3.55 +03	2.78 +03	2.30 +03	1.88 +03	1.82 +03	1.60 +03
24 CR	9.59 +03	7.89 +03	6.12 +03	5.13 +03	4.24 +03	3.31 +03	2.74 +03	2.23 +03	2.18 +03	1.82 +03
25 MN	1.10 +03	9.09 +03	7.03 +03	5.88 +03	4.86 +03	3.79 +03	3.12 +03	2.54 +03	2.61 +03	2.18 +03
26 FE	1.32 +03	1.20 +04	9.29 +03	7.79 +03	5.82 +03	5.03 +03	4.15 +03	3.38 +03	2.90 +03	2.43 +03
27 CO	1.46 +03	1.21 +03	9.46 +03	7.05 +03	6.44 +03	5.26 +03	4.36 +03	3.40 +03	2.93 +03	2.46 +03
28 NI	1.39 +03	1.16 +03	9.03 +03	7.61 +03	6.32 +03	5.26 +03	5.02 +03	3.58 +03	3.09 +03	2.60 +03
29 CU	1.22 +03	1.01 +03	9.46 +03	8.02 +03	6.67 +03	6.05 +03	5.02 +03	4.12 +03	3.55 +03	2.99 +03
30 ZN	3.12 +03	2.73 +03	9.85 +03	8.02 +03	7.08 +03	6.54 +03	5.47 +03	4.49 +03	3.87 +03	3.26 +03
31 GA	3.40 +03	3.22 +03	2.51 +03	2.17 +03	7.48 +03	7.29 +03	6.13 +03	5.01 +03	4.31 +03	3.62 +03
32 GE	3.91 +03	3.22 +03	2.67 +03	2.11 +03	1.81 +03	6.54 +03	6.54 +03	5.72 +03	4.66 +03	3.92 +03
33 AS	4.13 +03	2.84 +03	2.11 +03	1.71 +03	1.34 +03	6.88 +03	6.88 +03	5.72 +03	4.94 +03	4.16 +03
34 SE	3.53 +03	4.09 +03	3.19 +03	2.68 +03	2.23 +03	1.75 +03	4.28 +03	5.36 +03	5.42 +03	4.61 +03
35 BR	2.95 +03	2.41 +03	1.98 +03	1.72 +03	1.48 +03	1.23 +03	1.06 +03	3.81 +03	4.36 +03	4.84 +03
36 KR	2.81 +03	2.97 +03	3.87 +03	3.25 +03	2.70 +03	2.12 +03	1.75 +03	1.43 +03	3.59 +03	4.16 +03
37 RB	6.02 +03	5.44 +03	4.23 +03	3.56 +03	2.95 +03	2.31 +03	1.91 +03	1.57 +03	1.34 +03	4.56 +03
38 SR	6.60 +03	4.97 +03	3.87 +03	3.56 +03	2.30 +03	1.83 +03	1.53 +03	1.57 +03	1.34 +03	1.11 +03
39 Y	4.96 +03	4.13 +03	4.23 +03	2.76 +03	2.30 +03	1.83 +03	1.53 +03	1.27 +03	1.10 +03	9.32 +02
40 ZR	5.18 +03	4.30 +03	3.37 +03	2.85 +03	2.37 +03	1.88 +03	1.56 +03	1.29 +03	1.11 +03	9.44 +02

Z	El										
41	NB	5.74e+03	4.76e+03	3.73e+03	3.15e+03	2.62e+03	2.07e+03	1.72e+03	1.42e+03	1.23e+03	1.03e+03
42	MO	6.06e+03	5.05e+03	4.01e+03	3.40e+03	2.85e+03	2.28e+03	1.91e+03	1.59e+03	1.38e+03	1.17e+03
43	TC	6.42e+03	5.37e+03	4.27e+03	3.63e+03	3.05e+03	2.45e+03	2.06e+03	1.71e+03	1.49e+03	1.27e+03
44	RU	6.89e+03	5.76e+03	4.59e+03	3.91e+03	3.29e+03	2.64e+03	2.22e+03	1.85e+03	1.61e+03	1.37e+03
45	RH	7.40e+03	6.20e+03	4.94e+03	4.22e+03	3.55e+03	2.86e+03	2.50e+03	2.01e+03	1.75e+03	1.49e+03
46	PD	7.53e+03	6.34e+03	5.22e+03	4.36e+03	3.69e+03	2.98e+03	2.53e+03	2.08e+03	1.81e+03	1.54e+03
47	AG	7.94e+03	7.50e+03	5.09e+03	4.44e+03	3.73e+03	3.14e+03	2.53e+03	2.13e+03	1.86e+03	1.59e+03
48	CD	8.94e+03	7.94e+03	6.35e+03	5.42e+03	4.30e+03	3.69e+03	3.12e+03	2.61e+03	2.12e+03	1.81e+03
49	IN	9.46e+03	8.52e+03	6.81e+03	5.82e+03	4.58e+03	3.96e+03	3.34e+03	2.79e+03	2.44e+03	1.94e+03
50	SN	1.01e+04	9.04e+03	7.20e+03	6.01e+03	4.91e+03	4.14e+03	3.48e+03	2.92e+03	2.53e+03	2.08e+03
51	SB	1.25e+04	1.05e+04	7.26e+03	6.37e+03	5.16e+03	4.43e+03	3.91e+03	2.90e+03	2.53e+03	2.15e+03
52	TE	1.17e+04	1.27e+04	8.53e+03	7.01e+03	5.87e+03	4.68e+03	3.91e+03	3.24e+03	2.81e+03	2.38e+03
53	I	1.23e+04	1.03e+04	1.01e+04	7.27e+03	6.09e+03	4.87e+03	3.78e+03	3.19e+03	2.81e+03	2.42e+03
54	XE	1.43e+04	1.19e+04	9.91e+03	9.27e+03	5.42e+03	5.43e+03	3.99e+03	3.07e+03	3.50e+03	2.93e+03
55	CS	1.36e+04	1.23e+04	9.30e+03	8.97e+03	7.69e+03	6.04e+03	3.48e+03	4.23e+03	3.86e+03	3.05e+03
56	BA	1.49e+04	1.26e+04	1.01e+04	8.81e+03	7.95e+03	6.21e+03	5.08e+03	4.49e+03	4.11e+03	3.24e+03
57	LA	1.32e+04	1.12e+04	1.04e+04	9.62e+03	7.74e+03	6.54e+03	5.48e+03	4.77e+03	4.39e+03	3.46e+03
58	CE	4.58e+03	8.64e+03	9.91e+03	8.97e+03	8.19e+03	6.91e+03	5.82e+03	4.49e+03	4.80e+03	3.88e+03
59	PR	4.69e+03	8.15e+03	7.58e+03	8.34e+03	7.83e+03	6.72e+03	6.37e+03	5.32e+03	5.04e+03	4.27e+03
60	ND	4.98e+03	8.86e+03	1.11e+04	8.43e+03	7.95e+03	6.99e+03	6.15e+03	5.34e+03	5.24e+03	4.49e+03
61	PM	5.23e+03	5.44e+03	1.01e+04	9.62e+03	8.06e+03	7.32e+03	6.38e+03	5.66e+03	5.27e+03	4.55e+03
62	SM	5.84e+03	5.16e+03	1.04e+04	2.66e+03	8.30e+03	6.43e+03	6.76e+03	5.84e+03	4.86e+03	4.60e+03
63	EU	6.25e+03	5.99e+03	9.69e+03	3.02e+03	5.75e+03	6.85e+03	6.98e+03	5.72e+03	5.98e+03	4.83e+03
64	GD	6.54e+03	6.11e+03	1.01e+04	3.24e+03	2.46e+03	7.18e+03	6.40e+03	5.29e+03	5.14e+03	4.61e+03
65	TB	6.92e+03	6.32e+03	1.28e+04	3.63e+03	2.80e+03	7.53e+03	6.76e+03	5.61e+03	5.47e+03	4.32e+03
66	DY	7.21e+03	5.86e+03	2.52e+04	3.83e+03	3.03e+03	5.32e+03	6.98e+03	6.21e+03	5.05e+03	4.78e+03
67	HO	7.34e+03	5.51e+03	4.82e+04	3.08e+03	3.20e+03	2.19e+03	2.11e+03	4.36e+03	5.39e+03	4.61e+03
68	ER	7.59e+03	5.73e+03	4.99e+04	4.03e+03	3.41e+03	2.64e+03	2.27e+03	1.88e+03	5.75e+03	4.93e+03
69	TM	7.00e+03	5.60e+03	4.67e+04	3.95e+03	3.54e+03	2.72e+03	2.36e+03	1.95e+03	1.67e+03	5.09e+03
70	YB	7.18e+03	5.60e+03	4.56e+04	3.98e+03	3.36e+03	2.82e+03	2.27e+03	1.87e+03	1.58e+03	1.36e+03
71	LU	6.99e+03	6.04e+03	4.38e+04	3.74e+03	3.32e+03	2.66e+03	2.12e+03	1.90e+03	1.61e+03	1.41e+03
72	HF	6.84e+03	6.04e+03	4.61e+03	3.92e+03	3.29e+03	2.53e+03	2.22e+03	1.77e+03	1.62e+03	1.31e+03
73	TA	6.62e+03	7.48e+03	4.53e+03	3.89e+03	3.31e+03	2.63e+03	2.24e+03	1.85e+03	1.65e+03	1.37e+03
74	W	6.71e+03	5.60e+03	4.63e+03	3.99e+03	3.40e+03	2.70e+03	2.21e+03	1.94e+03	1.54e+03	1.37e+03
75	RE	6.84e+03	6.04e+03	4.56e+03	3.99e+03	3.63e+03	2.79e+03	2.30e+03	1.85e+03	1.61e+03	1.47e+03
76	OS	6.62e+03	5.80e+03	4.99e+03	5.39e+03	4.90e+03	2.94e+03	2.38e+03	1.94e+03	1.71e+03	1.53e+03
77	IR	6.71e+03	6.04e+03	6.51e+03	5.65e+03	4.72e+03	4.23e+03	3.41e+03	2.08e+03	1.78e+03	1.53e+03
78	PT	6.84e+03	7.48e+03	6.34e+03	1.75e+04	1.51e+04	1.22e+04	3.58e+03	2.83e+03	1.82e+03	1.37e+03
79	AU	6.71e+03	7.05e+03	2.17e+04					2.96e+03	2.45e+03	2.07e+03
80	HG	9.08e+03	2.85e+03					1.09e+04	8.97e+03	7.65e+03	6.33e+03
81	TL										
82	PB										
83	BI										
86	RN										
90	TH										
92	U										
94	PU	3.50e+03									

MASS ATTENUATION COEFFICIENTS FOR L ALPHA 1 LINES

EMITTER	ZR	NB	MO	TC	RU	RH	PD	AG	CD	IN
WAVELENGTH	6.07	5.72	5.40	5.11	4.84	4.59	4.36	4.15	3.95	3.77

ABSORBER

Z	El	ZR	NB	MO	TC	RU	RH	PD	AG	CD	IN
1	H	2.61+00	2.34+00	2.06+00	1.77+00	1.55+00	1.42+00	1.28+00	1.14+00	1.06+00	9.90-01
2	HE	1.04+01	9.13+00	7.79+00	6.40+00	5.33+00	4.70+00	4.05+00	3.38+00	3.00+00	2.64+00
3	LI	3.04+01	2.65+01	2.26+01	1.85+01	1.53+01	1.34+01	1.15+01	9.56+00	8.42+00	7.37+00
4	BE	7.36+01	6.43+01	5.47+01	4.49+01	3.72+01	3.26+01	2.79+01	2.31+01	2.03+01	1.77+01
5	B	1.49+02	1.30+02	1.11+02	9.17+01	7.63+01	6.70+01	5.74+01	4.76+01	4.19+01	3.66+01
6	C	2.85+02	2.49+02	2.13+02	1.74+02	1.45+02	1.27+02	1.08+02	8.99+01	7.90+01	6.88+01
7	N	4.56+02	3.99+02	3.41+02	2.82+02	2.35+02	2.06+02	1.77+02	1.47+02	1.30+02	1.13+02
8	O	6.80+02	5.98+02	5.13+02	4.25+02	3.57+02	3.14+02	2.70+02	2.25+02	1.99+02	1.74+02
9	F	8.33+02	7.31+02	6.27+02	5.20+02	4.36+02	3.84+02	3.30+02	2.76+02	2.43+02	2.13+02
10	NE	1.21+03	1.06+03	9.17+02	7.65+02	6.44+02	5.68+02	4.91+02	4.11+02	3.64+02	3.20+02
11	NA	1.48+03	1.30+03	1.12+03	9.34+02	7.86+02	6.94+02	5.99+02	5.01+02	4.44+02	3.90+02
12	MG	1.87+03	1.66+03	1.43+03	1.22+03	1.04+03	9.30+02	8.11+02	6.88+02	6.12+02	5.41+02
13	AL	2.25+03	2.03+03	1.74+03	1.47+03	1.26+03	1.12+03	9.81+02	8.34+02	7.43+02	6.58+02
14	SI	3.03+03	2.73+03	2.35+03	1.97+03	1.66+03	1.46+03	1.27+03	1.06+03	9.44+02	8.31+02
15	P	3.03+03	3.22+03	2.81+03	2.38+03	2.03+03	1.80+03	1.55+03	1.30+03	1.15+03	1.01+03
16	S	4.62+03	3.64+03	3.17+03	2.69+03	2.51+03	2.22+03	1.92+03	1.61+03	1.43+03	1.26+03
17	CL	4.10+03	4.10+03	3.89+03	3.03+03	2.56+03	2.25+03	2.17+03	1.86+03	1.65+03	1.45+03
18	AR	5.04+03	4.47+03	3.89+03	3.29+03	2.81+03	2.49+03	2.86+03	2.44+03	2.16+03	1.89+03
19	K	6.64+03	5.89+03	5.11+03	4.31+03	3.68+03	3.28+03	2.86+03	2.91+03	2.60+03	2.31+03
20	CA	8.03+03	7.11+03	6.16+03	5.19+03	4.42+03	3.93+03	3.42+03	3.14+03	2.80+03	2.74+03
21	SC	8.80+03	7.78+03	6.73+03	5.65+03	4.79+03	4.26+03	3.71+03	3.48+03	3.10+03	3.17+03
22	TI	9.95+03	8.78+03	7.58+03	6.34+03	5.36+03	4.75+03	4.12+03	3.92+03	3.48+03	3.58+03
23	V	1.13+04	1.00+04	8.63+03	7.20+03	6.07+03	5.37+03	4.65+03	4.58+03	4.06+03	4.02+02
24	CR	1.34+04	1.18+04	1.01+04	8.46+03	7.12+03	6.30+03	5.45+03	5.15+03	4.56+03	4.78+02
25	MN	1.52+04	1.59+03	1.37+03	9.57+03	8.04+03	7.10+03	6.14+03	6.12+02	5.42+02	5.38+02
26	FE	1.81+03	1.78+03	1.53+03	1.27+03	9.57+03	8.45+03	7.30+03	6.88+02	6.10+02	5.73+02
27	CO	2.02+03	1.83+03	1.58+03	1.32+03	1.07+03	9.48+03	8.20+03	7.28+02	6.48+02	6.06+02
28	NI	2.07+03	2.03+03	1.66+03	1.39+03	1.11+03	9.92+03	8.62+03	7.69+02	6.85+02	6.93+02
29	CU	2.18+03	1.93+03	1.91+03	1.60+03	1.18+03	1.04+03	9.10+03	8.80+02	7.84+02	7.60+02
30	ZN	2.51+03	2.21+03	1.91+03	1.75+03	1.35+03	1.31+03	1.04+03	9.65+02	8.60+02	8.23+02
31	GA	2.74+03	2.42+03	2.09+03	1.92+03	1.48+03	1.43+03	1.24+03	1.04+03	9.32+02	9.07+02
32	GE	3.03+03	2.67+03	2.30+03	2.09+03	1.62+03	1.57+03	1.36+03	1.15+03	1.02+03	9.70+02
33	AS	3.29+03	2.90+03	2.50+03	2.23+03	1.75+03	1.67+03	1.45+03	1.21+03	1.09+03	1.07+03
34	SE	3.49+03	3.08+03	2.66+03	2.49+03	1.89+03	1.86+03	1.62+03	1.36+03	1.21+03	1.14+03
35	BR	3.91+03	3.45+03	2.98+03	2.65+03	2.10+03	1.99+03	1.72+03	1.45+03	1.29+03	1.27+03
36	KR	4.18+03	3.68+03	3.18+03	2.92+03	2.24+03	2.21+03	1.92+03	1.62+03	1.44+03	1.38+03
37	RB	3.84+03	4.03+03	3.50+03	3.16+03	2.59+03	2.41+03	2.09+03	1.76+03	1.56+03	1.51+03
38	SR	4.21+03	3.58+03	3.69+03	3.43+03	2.73+03	2.65+03	2.29+03	1.93+03	1.71+03	1.60+03
39	Y	7.77+02	3.91+03	3.36+03	3.03+03	2.99+03	2.99+03	2.41+03	2.04+03	1.81+03	
40	ZR	7.89+02	6.76+02	2.34+03	3.03+03	3.13+03	2.77+03	2.41+03	2.04+03	1.81+03	

Z	El	c1	c2	c3	c4	c5	c6	c7	c8	c9	c10
41	NB	8.70+02	7.59+02	6.45+02	2.24+02	2.91+02	2.50+02	2.64+02	2.27+02	2.02+02	1.78+03
42	MO	9.98+02	8.86+02	7.71+02	6.53+02	2.02+02	2.72+02	2.35+02	2.44+02	2.18+02	1.92+02
43	TC	1.08+03	9.61+02	8.37+02	7.10+02	2.03+02	5.84+02	2.40+02	2.08+02	2.21+02	1.96+02
44	RU	1.17+03	1.13+03	9.21+02	7.70+02	6.58+02	6.48+02	5.70+02	2.21+02	1.94+02	1.42+02
45	RH	1.27+03	1.22+03	1.07+03	8.44+02	7.25+02	6.54+02	6.26+02	4.89+02	1.32+02	4.26+02
46	PD	1.37+03	1.38+03	1.20+03	9.16+02	7.91+02	7.09+02	6.92+02	5.40+02	4.83+02	4.77+02
47	AG	1.55+03	1.58+03	1.38+03	1.02+03	8.81+02	7.88+02	7.48+02	5.94+02	5.34+02	5.17+02
48	CD	1.66+03	1.63+03	1.42+03	1.18+03	1.03+03	8.50+02	7.99+02	6.43+02	5.78+02	5.52+02
49	IN	1.78+03	1.87+03	1.63+03	1.20+03	1.03+03	9.09+02	8.13+02	6.87+02	6.18+02	5.92+02
50	SN	2.02+03	2.15+03	1.85+03	1.38+03	1.17+03	9.26+02	8.13+02	6.96+02	6.25+02	5.98+02
51	SB	2.11+03	2.25+03	1.93+03	1.42+03	1.23+03	9.99+02	9.05+02	7.44+02	6.66+02	6.25+02
52	TE	2.44+03	2.40+03	2.21+03	1.63+03	1.36+03	1.05+03	9.31+02	7.84+02	7.02+02	6.51+02
53	I	2.55+03	2.56+03	2.37+03	1.61+03	1.56+03	1.14+03	1.04+03	8.33+02	7.39+02	6.82+02
54	XE	2.72+03	2.74+03	2.96+03	1.85+03	1.68+03	1.20+03	1.20+03	8.78+02	7.80+02	7.00+02
55	CS	2.90+03	2.88+03	2.21+03	1.98+03	1.87+03	1.29+03	1.29+03	9.44+02	8.05+02	7.42+02
56	BA	3.10+03	3.04+03	2.49+03	2.21+03	1.97+03	1.38+03	1.36+03	1.01+03	9.76+02	8.13+02
57	LA	3.26+03	3.19+03	2.63+03	2.32+03	2.18+03	1.49+03	1.44+03	1.15+03	1.03+03	8.63+02
58	CE	3.44+03	3.39+03	2.76+03	2.47+03	2.38+03	1.57+03	1.52+03	1.22+03	1.09+03	9.69+02
59	PR	3.61+03	3.52+03	2.94+03	2.72+03	2.50+03	1.66+03	1.69+03	1.29+03	1.15+03	1.02+03
60	ND	3.85+03	3.73+03	3.05+03	2.78+03	2.64+03	1.86+03	1.85+03	1.37+03	1.28+03	1.09+03
61	PM	3.98+03	3.86+03	3.23+03	2.90+03	2.79+03	1.94+03	1.95+03	1.43+03	1.35+03	1.14+03
62	SM	4.07+03	3.68+03	3.27+03	3.04+03	2.91+03	2.05+03	2.17+03	1.52+03	1.40+03	1.25+03
63	EU	4.12+03	3.91+03	3.39+03	3.19+03	3.02+03	2.12+03	2.26+03	1.66+03	1.56+03	1.31+03
64	GD	4.56+03	3.95+03	3.34+03	3.31+03	2.89+03	2.23+03	2.38+03	1.84+03	1.65+03	1.46+03
65	TB	4.33+03	3.83+03	3.52+03	3.30+03	3.02+03	2.48+03	2.46+03	1.92+03	1.72+03	1.52+03
66	DY	3.99+03	4.08+03	3.70+03	3.42+03	2.98+03	2.59+03	2.54+03	2.03+03	1.81+03	1.61+03
67	HO	4.07+03	3.86+03	3.79+03	3.18+03	3.08+03	2.73+03	2.62+03	2.11+03	1.89+03	1.68+03
68	ER	4.22+03	3.91+03	3.37+03	3.42+03	3.02+03	2.80+03	2.50+03	2.20+03	1.97+03	1.75+03
69	TM	4.56+03	3.68+03	3.45+03	2.95+03	2.78+03	2.61+03	2.46+03	2.39+03	2.06+03	1.86+03
70	YB	4.33+03	3.95+03	3.52+03	2.40+03	2.83+03	2.84+03	2.33+03	2.13+03	2.09+03	1.91+03
71	LU	3.99+03	4.08+03	3.70+03	2.67+03	2.98+03	2.55+03	2.19+03	2.15+03	1.96+03	1.87+03
72	HF	3.17+03	3.12+03	3.71+03	3.17+03	3.08+03	2.21+03	2.34+03	1.72+03	1.68+03	1.49+03
73	TA	4.52+03	3.56+03	3.37+03	2.15+03	2.78+03	2.45+03	2.34+03	2.21+03	1.97+03	1.89+03
74	W	4.33+03	3.95+03	3.45+03	9.07+02	2.83+03	2.51+03	2.34+03	2.25+03	2.01+03	1.78+03
75	RE	3.01+03	3.83+03	3.51+03	8.93+02	2.08+03	2.67+03	2.34+03	2.06+03	1.82+03	1.94+03
76	OS	1.16+03	2.08+03	2.75+02	3.17+03	2.78+03	6.84+02	5.98+02	6.48+02	1.74+03	1.56+03
77	IR	1.15+03	1.03+03	8.55+02	2.15+03	2.83+03	8.70+02	7.61+02	6.65+02	5.94+02	5.28+02
78	PT	1.16+03	1.13+03	8.87+02	2.07+03	2.00+03	8.96+02	7.82+02	1.66+03	1.46+03	1.28+03
79	AU	1.32+03	1.19+03	1.05+03	1.18+03	7.81+02	2.33+03	2.00+03	1.66+03	1.46+03	1.28+03
80	HG	1.26+03	1.56+03	1.04+03	1.40+03	1.69+03					
81	TL	1.33+03	1.62+03	1.35+03		1.00+03					
82	PB	1.76+03	1.83+03	1.40+03		2.66+03					
83	BI	1.26+03	4.57+03	3.89+03	3.20+03						
90	TH										
92	U	5.22+03									
94	PU										

MASS ATTENUATION COEFFICIENTS FOR L ALPHA 1 LINES

EMITTER	SN	SB	TE	I	XE	CS	BA	LA	CE	PR
WAVELENGTH	3.59	3.43	3.28	3.14	3.01	2.89	2.77	2.66	2.56	2.46

ABSORBER										
1 H	9.12 -01	8.55 -01	8.10 -01	7.63 -01	7.30 -01	7.04 -01	6.78 -01	6.51 -01	6.23 -01	5.98 -01
2 HE	2.28 +00	2.01 +00	1.80 +00	1.58 +00	1.44 +00	1.32 +00	1.21 +00	1.09 +00	9.76 -01	8.65 -01
3 LI	6.28 +00	5.49 +00	4.86 +00	4.22 +00	3.77 +00	3.44 +00	3.11 +00	2.75 +00	2.40 +00	2.07 +00
4 BE	1.51 +01	1.32 +01	1.16 +01	1.00 +01	8.97 +00	8.16 +00	7.33 +00	6.48 +00	5.61 +00	4.81 +00
5 B	3.12 +01	2.72 +01	2.41 +01	2.15 +01	1.85 +01	1.68 +01	1.51 +01	1.33 +01	1.15 +01	9.90 +00
6 C	5.84 +01	5.08 +01	4.47 +01	3.85 +01	3.42 +01	3.10 +01	2.78 +01	2.44 +01	2.10 +01	1.79 +01
7 N	9.70 +01	8.47 +01	7.48 +01	6.47 +01	5.77 +01	5.24 +01	4.70 +01	4.15 +01	3.58 +01	3.06 +01
8 O	1.48 +02	1.30 +02	1.14 +02	9.95 +01	8.87 +01	8.06 +01	7.23 +01	6.38 +01	5.51 +01	4.71 +01
9 F	1.83 +02	1.50 +02	1.41 +02	1.23 +02	1.09 +02	9.99 +01	8.98 +01	7.95 +01	6.88 +01	5.90 +01
10 NE	2.74 +02	2.41 +02	2.13 +02	1.85 +02	1.66 +02	1.51 +02	1.36 +02	1.20 +02	1.04 +02	8.97 +01
11 NA	3.34 +02	2.94 +02	2.60 +02	2.26 +02	2.02 +02	1.84 +02	1.66 +02	1.47 +02	1.27 +02	1.09 +02
12 MG	4.69 +02	4.14 +02	3.68 +02	3.22 +02	2.89 +02	2.64 +02	2.38 +02	2.12 +02	1.84 +02	1.59 +02
13 AL	5.70 +02	5.04 +02	4.49 +02	3.93 +02	3.53 +02	3.23 +02	2.92 +02	2.59 +02	2.26 +02	1.96 +02
14 SI	7.14 +02	6.27 +02	5.56 +02	4.84 +02	4.33 +02	3.94 +02	3.55 +02	3.14 +02	2.73 +02	2.35 +02
15 P	8.69 +02	7.63 +02	6.77 +02	5.88 +02	5.26 +02	4.79 +02	4.31 +02	3.82 +02	3.31 +02	2.85 +02
16 S	1.08 +03	9.56 +02	8.49 +02	7.39 +02	6.62 +02	6.03 +02	5.44 +02	4.82 +02	4.19 +02	3.61 +02
17 CL	1.25 +03	1.10 +03	9.80 +02	8.53 +02	7.64 +02	6.96 +02	6.27 +02	5.56 +02	4.83 +02	4.16 +02
18 AR	1.37 +03	1.22 +03	1.08 +03	9.51 +02	8.54 +02	7.80 +02	7.04 +02	6.26 +02	5.46 +02	4.72 +02
19 K	1.68 +03	1.48 +03	1.57 +03	1.40 +03	1.10 +03	1.00 +03	9.04 +02	7.99 +02	6.92 +02	5.94 +02
20 CA	2.01 +03	1.76 +03	1.57 +03	1.40 +03	1.34 +03	1.22 +03	1.10 +03	9.75 +02	8.45 +02	7.25 +02
21 SC	2.15 +03	1.91 +03	1.71 +03	1.50 +03	1.50 +03	1.31 +03	1.16 +03	1.04 +03	8.84 +02	7.80 +02
22 TI	2.37 +02	2.09 +02	1.87 +02	1.82 +02	1.64 +02	1.50 +02	1.35 +02	1.20 +02	1.05 +02	9.22 +01
23 V	2.65 +02	2.34 +02	2.08 +02	2.12 +02	1.90 +02	1.74 +02	1.57 +02	1.40 +02	1.22 +02	1.06 +02
24 CR	3.09 +02	2.72 +02	2.42 +02	2.37 +02	2.12 +02	1.94 +02	1.75 +02	1.56 +02	1.36 +02	1.18 +02
25 MN	3.46 +02	3.05 +02	2.71 +02	2.81 +02	2.52 +02	2.30 +02	2.08 +02	1.85 +02	1.61 +02	1.39 +02
26 FE	4.11 +02	3.62 +02	3.22 +02	3.17 +02	2.85 +02	2.60 +02	2.35 +02	2.09 +02	1.83 +02	1.58 +02
27 CO	4.63 +02	4.08 +02	3.63 +02	3.44 +02	3.10 +02	2.84 +02	2.57 +02	2.30 +02	2.02 +02	1.76 +02
28 NI	4.96 +02	4.39 +02	3.92 +02	3.64 +02	3.28 +02	3.00 +02	2.72 +02	2.43 +02	2.14 +02	1.86 +02
29 CU	5.25 +02	4.64 +02	4.15 +02	4.16 +02	3.74 +02	3.43 +02	3.12 +02	2.77 +02	2.43 +02	2.12 +02
30 ZN	6.00 +02	5.31 +02	4.74 +02	4.57 +02	4.12 +02	3.77 +02	3.42 +02	3.05 +02	2.68 +02	2.34 +02
31 GA	6.59 +02	5.83 +02	5.20 +02	4.94 +02	4.41 +02	4.03 +02	3.65 +02	3.25 +02	2.85 +02	2.47 +02
32 GE	7.11 +02	6.28 +02	5.59 +02	5.44 +02	4.90 +02	4.48 +02	4.06 +02	3.63 +02	3.18 +02	2.77 +02
33 AS	7.85 +02	6.94 +02	6.20 +02	5.82 +02	5.25 +02	4.80 +02	4.35 +02	3.89 +02	3.42 +02	2.98 +02
34 SE	8.40 +02	7.43 +02	6.64 +02	6.43 +02	5.79 +02	5.29 +02	4.80 +02	4.28 +02	3.75 +02	3.27 +02
35 BR	9.30 +02	8.22 +02	7.33 +02	6.85 +02	6.16 +02	5.64 +02	5.11 +02	4.56 +02	4.00 +02	3.48 +02
36 KR	9.91 +02	8.76 +02	7.81 +02	7.56 +02	6.80 +02	6.22 +02	5.63 +02	5.02 +02	4.40 +02	3.82 +02
37 RB	1.09 +03	9.69 +02	8.64 +02	8.22 +02	7.39 +02	6.76 +02	6.11 +02	5.45 +02	4.77 +02	4.14 +02
38 SR	1.19 +03	1.05 +03	9.39 +02	8.99 +02	8.09 +02	7.39 +02	6.69 +02	5.96 +02	5.22 +02	4.53 +02
39 Y	1.30 +03	1.15 +03	1.02 +03	8.99 +02	8.60 +02	7.87 +02	7.12 +02	6.35 +02	5.57 +02	4.84 +02
40 ZR	1.38 +03	1.22 +03	1.09 +03	9.56 +02						

41	NB	1.54+03	1.36+03	1.21+03	1.06+03	9.54+02	8.72+02	7.89+02	7.03+02	6.16+02	5.35+02
42	MO	1.66+03	1.46+03	1.30+03	1.14+03	1.02+03	9.37+02	8.43+02	7.55+02	6.60+02	5.73+02
43	TC	1.70+03	1.50+03	1.34+03	1.17+03	1.05+03	9.65+02	8.73+02	7.79+02	6.82+02	5.92+02
44	RU	1.80+03	1.60+03	1.42+03	1.25+03	1.12+03	1.03+03	9.32+02	8.31+02	7.28+02	6.33+02
45	RH	1.78+03	1.63+03	1.37+03	1.25+03	1.08+03	9.98+02	9.05+02	8.11+02	7.13+02	6.23+02
46	PD	1.58+03	1.84+03	1.65+03	1.45+03	1.31+03	1.20+03	1.08+03	9.68+02	8.47+02	7.36+02
47	AG	1.11+03	1.30+03	1.25+03	1.50+03	1.36+03	1.24+03	1.18+03	1.00+03	8.83+02	7.68+02
48	CD	1.19+02	1.13+03	1.41+03	1.04+03	1.29+03	1.31+03	1.20+03	1.06+03	9.30+02	8.09+02
49	IN	4.54+02	4.03+02	3.89+02	9.54+02	3.10+02	1.06+03	9.73+02	1.10+03	9.65+02	8.51+02
50	SN	4.85+02	4.33+02	3.93+02	3.48+02	3.25+02	7.86+02	7.43+02	8.96+02	9.87+02	8.86+02
51	SB	5.16+02	4.36+02	4.12+02	3.64+02	3.46+02	2.91+02	3.41+02	9.12+02	9.56+02	8.90+02
52	TE	4.88+02	4.86+02	4.00+02	3.86+02	3.07+02	3.13+02	2.80+02	7.39+02	6.61+02	9.64+02
53	I	5.46+02	5.51+02	4.41+02	4.47+02	3.68+02	3.74+02	3.05+02	3.07+02	6.85+02	6.15+02
54	XE	5.95+02	5.25+02	5.06+02	3.85+02	3.99+02	3.16+02	3.41+02	2.72+02	2.23+02	2.06+02
55	CS	5.62+02	5.67+02	4.68+02	4.43+02	3.46+02	3.65+02	2.81+02	2.95+02	2.38+02	2.25+02
56	BA	6.12+02	6.12+02	5.06+02	5.19+02	3.99+02	3.59+02	3.05+02	3.48+02	2.59+02	2.66+02
57	LA	6.41+02	6.62+02	5.47+02	5.51+02	4.32+02	4.29+02	3.30+02	3.20+02	3.06+02	2.84+02
58	GE	6.93+02	7.02+02	6.27+02	6.20+02	4.97+02	4.85+02	3.59+02	3.48+02	3.25+02	3.03+02
59	PR	7.48+02	7.46+02	7.04+02	6.62+02	5.29+02	5.13+02	3.89+02	3.70+02	3.47+02	3.21+02
60	ND	7.92+02	7.87+02	7.52+02	7.32+02	5.59+02	5.48+02	4.13+02	3.94+02	3.67+02	3.43+02
61	PM	8.41+02	8.40+02	7.88+02	7.66+02	5.97+02	5.76+02	4.13+02	4.17+02	3.47+02	3.62+02
62	SM	8.88+02	8.80+02	8.32+02	8.08+02	6.61+02	6.07+02	4.66+02	4.46+02	4.35+02	3.81+02
63	EU	9.91+02	9.29+02	8.68+02	8.50+02	6.93+02	6.36+02	5.23+02	4.94+02	4.58+02	4.24+02
64	GD	9.47+02	9.68+02	9.64+02	8.99+02	7.69+02	6.71+02	5.51+02	5.47+02	4.83+02	4.46+02
65	TB	1.04+03	1.07+03	1.06+03	9.35+02	8.13+02	7.06+02	5.78+02	5.76+02	5.09+02	4.71+02
66	DY	1.14+03	1.13+03	1.16+03	9.82+02	8.45+02	7.46+02	6.10+02	6.09+02	5.38+02	4.89+02
67	HO	1.21+03	1.18+03	1.20+03	1.05+03	8.88+02	7.76+02	6.42+02	6.33+02	5.58+02	5.13+02
68	ER	1.33+03	1.24+03	1.26+03	1.11+03	9.24+02	8.47+02	7.05+02	6.64+02	5.85+02	5.32+02
69	TM	1.40+03	1.29+03	1.31+03	1.17+03	9.62+02	8.82+02	7.40+02	6.90+02	6.08+02	5.54+02
70	YB	1.48+03	1.35+03	1.36+03	1.20+03	1.01+03	9.28+02	8.01+02	7.18+02	6.33+02	5.85+02
71	LU	1.43+03	1.41+03	1.31+03	1.29+03	1.05+03	9.73+02	8.43+02	7.57+02	7.01+02	6.15+02
72	HF	1.46+03	1.46+03	1.36+03	1.34+03	1.09+03	9.65+02	8.85+02	7.94+02	6.39+02	6.14+02
73	TA	1.52+03	1.52+03	1.46+03	1.43+03	1.17+03	1.08+03	9.13+02	8.20+02	7.62+02	5.87+02
74	W	1.59+03	1.54+03	1.49+03	1.18+03	1.21+03	9.98+02	9.82+02	7.43+02	7.83+02	6.95+02
75	RE	1.66+03	1.62+03	1.46+03	1.21+03	1.34+03	1.07+03	8.97+02	8.84+02	7.92+02	7.10+02
76	OS	1.63+03	1.52+03	1.50+03	1.34+03	1.06+03	1.11+03	1.09+03	8.96+02	8.09+02	7.65+02
77	IR	1.70+03	1.64+03	1.50+03	1.31+03	1.09+03	1.23+03	1.01+03	9.84+02	8.71+02	5.87+02
78	PT	1.47+03	1.47+03	1.33+03	1.34+03	1.05+03	1.05+03	1.12+03	1.01+03	8.96+02	6.95+02
79	AU	1.60+03	1.51+03	1.18+03	1.18+03	1.06+03	9.59+02	9.51+02	9.55+02	8.57+02	7.88+02
80	HG	1.6R+03	1.64+03	1.21+03	1.21+03	2.40+02	9.59+02	1.05+03	9.84+02	8.70+02	8.70+02
81	TL	1.71+03	1.47+03	1.46+02	1.80+02	9.36+02	2.14+02	1.87+03	1.97+03	1.76+03	7.78+02
82	PB	1.68+03	1.51+03	1.49+02							1.55+03
83	BI	1.72+03	1.64+02	1.46+02							
86	RN	1.37+03	1.47+02	1.33+02							
90	TH	1.12+03	1.12+02	1.31+02							
92	U	4.60+02	9.77+02	1.31+02							
94	PU	1.09+03	9.48+02	8.25+02							

MASS ATTENUATION COEFFICIENTS FOR L ALPHA 1 LINES

EMITTER	ND	PM	SM	EU	GD	TB	DY	HO	ER	TM
WAVELENGTH	2.37	2.28	2.19	2.12	2.04	1.97	1.90	1.84	1.78	1.72

ABSORBER

		ND	PM	SM	EU	GD	TB	DY	HO	ER	TM
1	H	5.83 -01	5.69 -01	5.54 -01	5.39 -01	5.26 -01	5.17 -01	5.07 -01	4.98 -01	4.88 -01	4.81 -01
2	HE	8.10 -01	7.53 -01	6.96 -01	6.37 -01	5.46 -01	5.54 -01	5.22 -01	4.90 -01	4.56 -01	4.35 -01
3	LI	1.91 +00	1.74 +00	1.58 +00	1.41 +00	1.26 +00	1.17 +00	1.08 +00	9.86 -01	8.91 -01	8.29 -01
4	BE	4.41 +00	4.00 +00	3.60 +00	3.17 +00	2.81 +00	2.59 +00	2.37 +00	2.14 +00	1.91 +00	1.75 +00
5	B	9.07 +00	8.22 +00	7.37 +00	6.48 +00	5.71 +00	5.25 +00	4.78 +00	4.30 +00	3.82 +00	3.50 +00
6	C	1.63 +01	1.47 +01	1.31 +01	1.15 +01	1.01 +01	9.26 +00	8.40 +00	7.52 +00	6.63 +00	6.04 +00
7	N	2.80 +01	2.54 +01	2.27 +01	1.99 +01	1.75 +01	1.61 +01	1.46 +01	1.31 +01	1.16 +01	1.05 +01
8	O	4.31 +01	3.89 +01	3.48 +01	3.05 +01	2.68 +01	2.45 +01	2.23 +01	1.99 +01	1.76 +01	1.60 +01
9	F	5.41 +01	4.90 +01	4.39 +01	3.86 +01	3.40 +01	3.12 +01	2.84 +01	2.55 +01	2.25 +01	2.06 +01
10	NE	8.73 +01	7.46 +01	6.68 +01	5.86 +01	5.19 +01	4.77 +01	4.33 +01	3.89 +01	3.44 +01	3.15 +01
11	NA	1.00 +02	9.12 +01	8.18 +01	7.20 +01	6.35 +01	5.84 +01	5.31 +01	4.77 +01	4.23 +01	3.86 +01
12	MG	1.46 +02	1.33 +02	1.19 +02	1.05 +02	9.33 +01	8.58 +01	7.80 +01	7.02 +01	6.24 +01	5.69 +01
13	AL	1.80 +02	1.64 +02	1.47 +02	1.30 +02	1.15 +02	1.06 +02	9.63 +01	8.72 +01	7.74 +01	7.09 +01
14	SI	2.15 +02	1.95 +02	1.75 +02	1.54 +02	1.36 +02	1.25 +02	1.13 +02	1.02 +02	9.05 +01	8.27 +01
15	P	2.61 +02	2.37 +02	2.12 +02	1.87 +02	1.65 +02	1.52 +02	1.38 +02	1.24 +02	1.10 +02	1.00 +02
16	S	3.31 +02	3.01 +02	2.70 +02	2.38 +02	2.10 +02	1.93 +02	1.76 +02	1.58 +02	1.40 +02	1.28 +02
17	CL	3.82 +02	3.46 +02	3.11 +02	2.74 +02	2.42 +02	2.22 +02	2.02 +02	1.82 +02	1.61 +02	1.47 +02
18	AR	4.33 +02	3.94 +02	3.54 +02	3.13 +02	2.77 +02	2.55 +02	2.32 +02	2.09 +02	1.86 +02	1.70 +02
19	K	5.44 +02	4.93 +02	4.41 +02	3.88 +02	3.42 +02	3.14 +02	2.85 +02	2.56 +02	2.27 +02	2.07 +02
20	CA	6.64 +02	6.01 +02	5.38 +02	4.72 +02	4.15 +02	3.81 +02	3.46 +02	3.11 +02	2.74 +02	2.50 +02
21	SC	7.15 +02	6.48 +02	5.81 +02	5.11 +02	4.50 +02	4.14 +02	3.76 +02	3.38 +02	2.99 +02	2.73 +02
22	TI	8.36 +02	6.10 +02	5.54 +02	4.96 +02	4.45 +02	4.12 +02	3.78 +02	3.44 +02	3.08 +02	2.84 +02
23	V	8.36 +02	7.48 +02	5.90 +02	5.36 +02	4.86 +02	4.49 +02	4.12 +02	3.74 +02	3.35 +02	3.08 +02
24	CR	9.61 +02	8.59 +02	7.84 +02	7.07 +02	6.10 +02	5.63 +02	5.14 +02	4.65 +02	4.16 +02	3.82 +02
25	MN	1.08 +02	9.86 +01	8.85 +01	7.92 +01	8.31 +01	6.43 +02	5.80 +02	5.10 +02	4.63 +02	4.26 +02
26	FE	1.28 +02	1.17 +02	1.05 +02	1.06 +02	9.48 +01	8.74 +01	5.85 +02	6.26 +02	5.67 +02	5.20 +01
27	CO	1.46 +02	1.33 +02	1.20 +02	1.20 +02	1.07 +02	9.91 +01	7.98 +01	7.21 +01	6.44 +02	5.93 +01
28	NI	1.62 +02	1.48 +02	1.34 +02	1.27 +02	1.13 +02	1.04 +02	9.62 +01	8.74 +01	7.41 +01	6.78 +01
29	CU	1.72 +02	1.57 +02	1.42 +02	1.44 +02	1.29 +02	1.19 +02	1.09 +02	9.93 +01	7.84 +01	7.23 +01
30	ZN	1.96 +02	1.79 +02	1.62 +02	1.59 +02	1.42 +02	1.31 +02	1.20 +02	1.14 +02	8.91 +01	8.21 +01
31	GA	2.15 +02	1.97 +02	1.78 +02	1.67 +02	1.49 +02	1.38 +02	1.26 +02	1.29 +02	9.83 +01	9.06 +01
32	GE	2.28 +02	2.08 +02	1.88 +02	1.67 +02	1.49 +02	1.55 +02	1.42 +02	1.14 +02	1.02 +02	9.44 +01
33	AS	2.55 +02	2.33 +02	2.11 +02	1.88 +02	1.81 +02	1.67 +02	1.53 +02	1.39 +02	1.16 +02	1.07 +02
34	SE	2.74 +02	2.51 +02	2.27 +02	2.02 +02	1.97 +02	1.83 +02	1.67 +02	1.52 +02	1.25 +02	1.25 +02
35	BR	3.01 +02	2.75 +02	2.49 +02	2.21 +02	2.10 +02	1.94 +02	1.78 +02	1.61 +02	1.36 +02	1.33 +02
36	KR	3.21 +02	2.93 +02	2.65 +02	2.36 +02	2.30 +02	2.12 +02	1.94 +02	1.76 +02	1.45 +02	1.45 +02
37	RB	3.52 +02	3.21 +02	2.90 +02	2.58 +02	2.49 +02	2.30 +02	2.10 +02	1.91 +02	1.58 +02	1.57 +02
38	SR	3.17 +02	3.48 +02	3.14 +02	2.79 +02	2.72 +02	2.51 +02	2.30 +02	2.08 +02	1.71 +02	1.71 +02
39	Y	4.17 +02	3.50 +02	3.43 +02	3.05 +02	2.72 +02	2.51 +02	2.30 +02	2.08 +02	1.86 +02	1.86 +02
40	ZR	4.46 +02	4.07 +02	3.67 +02	3.27 +02	2.91 +02	2.69 +02	2.46 +02	2.23 +02	2.00 +02	1.84 +02

41	NB	4.92+02	4.49+02	4.05+02	3.60+02	3.21+02	2.96+02	2.71+02	2.46+02	2.20+02	2.02+02
42	MO	5.27+02	4.81+02	4.34+02	3.85+02	3.43+02	3.16+02	2.89+02	2.63+02	2.34+02	2.15+02
43	TC	5.46+02	4.98+02	4.49+02	4.00+02	3.56+02	3.29+02	3.01+02	2.73+02	2.44+02	2.25+02
44	RU	5.83+02	5.32+02	4.81+02	4.28+02	3.81+02	3.52+02	3.23+02	2.92+02	2.62+02	2.41+02
45	RH	6.78+02	5.76+02	5.26+02	4.77+02	4.26+02	3.81+02	3.52+02	3.23+02	2.95+02	2.65+02
46	PD	6.78+02	6.18+02	5.58+02	4.96+02	4.42+02	4.08+02	3.74+02	3.38+02	3.03+02	2.78+02
47	AG	7.08+02	6.47+02	5.85+02	5.21+02	4.64+02	4.29+02	3.93+02	3.57+02	3.19+02	2.94+02
48	CD	7.46+02	6.81+02	6.18+02	5.47+02	4.87+02	4.51+02	4.12+02	3.74+02	3.35+02	3.08+02
49	IN	7.85+02	7.18+02	6.49+02	5.79+02	5.17+02	4.78+02	4.39+02	3.98+02	3.57+02	3.29+02
50	SN	8.17+02	7.46+02	6.74+02	6.01+02	5.36+02	4.96+02	4.54+02	4.12+02	3.69+02	3.40+02
51	SB	8.30+02	7.52+02	6.82+02	6.09+02	5.45+02	5.05+02	4.64+02	4.22+02	3.76+02	3.50+02
52	TE	8.90+02	8.14+02	7.37+02	6.58+02	5.88+02	5.45+02	5.00+02	4.54+02	4.08+02	3.76+02
53	I	9.65+02	8.86+02	8.21+02	7.23+02	6.49+02	6.01+02	5.50+02	5.00+02	4.48+02	4.12+02
54	XE	8.38+02	7.49+02	7.41+02	7.43+02	6.73+02	6.22+02	5.70+02	5.18+02	4.64+02	4.28+02
55	CS	6.07+02	8.18+02	7.85+02	8.03+02	7.32+02	6.77+02	6.20+02	5.63+02	5.04+02	4.64+02
56	BA	2.04+02	1.83+02	2.03+02	5.27+02	7.71+02	7.75+02	6.53+02	5.92+02	5.30+02	4.83+02
57	LA	2.46+02	2.24+02	2.17+02	5.48+02	7.13+02	7.12+02	6.58+02	6.00+02	5.61+02	5.39+02
58	CE	2.24+02	2.03+02	1.82+02	1.94+02	4.96+02	6.78+02	6.14+02	6.05+02	5.85+02	5.65+02
59	PR	2.80+02	2.40+02	2.32+02	2.07+02	1.73+02	1.69+02	4.83+02	6.54+02	6.11+02	5.95+02
60	ND	2.68+02	2.56+02	2.46+02	2.20+02	1.85+02	1.81+02	1.65+02	4.36+02	5.94+02	6.05+02
61	PM	2.96+02	2.71+02	2.63+02	2.35+02	1.97+02	1.95+02	1.79+02	1.63+02	1.47+02	4.24+02
62	SM	3.17+02	3.06+02	2.92+02	2.49+02	2.11+02	2.07+02	1.99+02	1.82+02	1.56+02	1.42+02
63	EU	3.52+02	3.22+02	3.09+02	2.61+02	2.34+02	2.17+02	2.12+02	1.93+02	1.63+02	1.50+02
64	GD	3.71+02	3.40+02	3.26+02	2.77+02	2.49+02	2.31+02	2.14+02	2.15+02	1.74+02	1.61+02
65	TB	3.92+02	3.59+02	3.44+02	2.92+02	2.76+02	2.43+02	2.36+02	2.27+02	1.84+02	1.70+02
66	DY	4.13+02	3.78+02	3.63+02	3.08+02	2.92+02	2.56+02	2.58+02	2.35+02	1.94+02	1.79+02
67	HO	4.36+02	3.99+02	3.76+02	3.25+02	3.16+02	2.71+02	2.69+02	2.45+02	2.04+02	1.89+02
68	ER	4.36+02	3.78+02	3.94+02	3.37+02	3.27+02	2.93+02	2.79+02	2.54+02	2.12+02	1.96+02
69	TM	4.54+02	4.14+02	3.88+02	3.53+02	3.41+02	3.03+02	2.90+02	2.64+02	2.21+02	2.04+02
70	YB	4.74+02	4.34+02	4.25+02	3.65+02	3.61+02	3.16+02	3.08+02	2.81+02	2.38+02	2.11+02
71	LU	4.92+02	4.50+02	4.50+02	3.80+02	3.83+02	3.35+02	3.27+02	2.96+02	2.53+02	2.20+02
72	HF	5.12+02	4.69+02	4.25+02	4.03+02	3.61+02	3.53+02	3.38+02	2.99+02	2.70+02	2.34+02
73	TA	5.41+02	4.95+02	4.50+02	4.26+02	3.80+02	3.55+02	3.69+02	3.08+02	2.78+02	2.46+02
74	W	5.68+02	5.22+02	4.74+02	4.41+02	3.96+02	3.67+02	3.78+02	3.37+02	3.04+02	2.49+02
75	RE	5.89+02	5.40+02	4.91+02	4.60+02	4.31+02	3.80+02	3.74+02	3.45+02	3.03+02	2.51+02
76	OS	5.46+02	5.03+02	4.60+02	4.79+02	4.32+02	3.76+02	4.08+02	3.72+02	3.36+02	2.82+02
77	IR	6.38+02	5.86+02	5.33+02	4.81+02	4.77+02	4.01+02	4.29+02	3.85+02	3.57+02	2.81+02
78	PT	6.43+02	5.90+02	5.36+02	5.31+02	4.93+02	4.10+02	4.08+02	3.93+02	3.36+02	3.11+02
79	AU	7.08+02	6.50+02	5.48+02	5.48+02	4.98+02	4.43+02	4.57+02	4.56+02	3.57+02	3.22+02
80	HG	7.30+02	6.69+02	5.91+02	5.51+02	5.31+02	4.57+02	4.64+02	4.36+02	4.33+02	3.31+02
81	TL	7.26+02	6.50+02	6.11+02	5.51+02	6.03+02	4.43+02	5.62+02	4.80+02	4.02+02	4.02+02
82	PB	6.73+02	6.07+02	6.11+02	6.71+02	6.10+02	4.57+02	5.24+02	5.19+02	4.77+02	6.59+02
83	BI	7.26+02	6.70+02	7.17+02	6.58+02	6.03+02	4.64+02	4.29+02	4.80+02	4.77+02	4.02+02
86	RN	8.07+02	7.80+02	7.22+02	6.56+02	6.03+02	5.62+02	5.24+02	4.80+02	4.33+02	6.59+02
90	TH	1.24+03	8.73+02	1.22+03	1.09+03	1.08+03	9.97+02	9.06+02	8.15+02	7.21+02	6.59+02
92	U	1.40+03	1.40+03	1.22+03	1.09+03	1.08+03	9.97+02	9.06+02	8.15+02	7.21+02	6.59+02
94	PU	1.40+03	1.24+03	1.22+03	1.09+03	1.08+03	9.97+02	9.06+02	8.15+02	7.21+02	6.59+02

MASS ATTENUATION COEFFICIENTS FOR L ALPHA 1 LINES

EMITTER	YB	LU	HF	TA	W	RE	OS	IR	PT	AU
WAVELENGTH	1.67	1.61	1.56	1.52	1.47	1.43	1.39	1.35	1.31	1.27

ABSORBER

1 H	4.76 -01	4.70 -01	4.64 -01	4.59 -01	4.54 -01	4.49 -01	4.44 -01	4.41 -01	4.37 -01	4.34 -01
2 HE	4.16 -01	3.97 -01	3.77 -01	3.62 -01	3.49 -01	3.36 -01	3.23 -01	3.13 -01	3.05 -01	2.96 -01
3 LI	7.75 -01	7.20 -01	6.65 -01	6.21 -01	5.86 -01	5.51 -01	5.14 -01	4.87 -01	4.63 -01	4.39 -01
4 BE	1.62 +00	1.49 +00	1.35 +00	1.25 +00	1.16 +00	1.08 +00	9.93 -01	9.26 -01	8.69 -01	8.11 -01
5 B	3.22 +00	2.94 +00	2.65 +00	2.42 +00	2.25 +00	2.07 +00	1.88 +00	1.74 +00	1.62 +00	1.50 +00
6 C	5.54 +00	5.03 +00	4.51 +00	4.10 +00	3.79 +00	3.46 +00	3.13 +00	2.88 +00	2.67 +00	2.45 +00
7 N	9.71 +00	8.82 +00	7.91 +00	7.21 +00	6.65 +00	6.08 +00	5.51 +00	5.06 +00	4.68 +00	4.30 +00
8 O	1.47 +01	1.33 +01	1.19 +01	1.08 +01	9.98 +00	9.11 +00	8.22 +00	7.54 +00	6.96 +00	6.37 +00
9 F	1.89 +01	1.72 +01	1.54 +01	1.40 +01	1.29 +01	1.18 +01	1.63 +01	9.87 +00	9.13 +00	8.38 +00
10 NE	2.89 +01	2.62 +01	2.35 +01	2.15 +01	1.98 +01	1.81 +01	1.63 +01	1.50 +01	1.39 +01	1.27 +01
11 NA	3.55 +01	3.22 +01	2.90 +01	2.64 +01	2.43 +01	2.23 +01	2.01 +01	1.85 +01	1.71 +01	1.57 +01
12 MG	5.22 +01	4.75 +01	4.26 +01	3.89 +01	3.58 +01	3.27 +01	2.96 +01	2.72 +01	2.51 +01	2.30 +01
13 AL	6.51 +01	5.92 +01	5.33 +01	4.86 +01	4.48 +01	4.10 +01	3.71 +01	3.41 +01	3.16 +01	2.89 +01
14 SI	7.59 +01	6.90 +01	6.19 +01	5.64 +01	5.20 +01	4.75 +01	4.30 +01	3.95 +01	3.65 +01	3.34 +01
15 P	9.23 +01	8.38 +01	7.53 +01	6.87 +01	6.33 +01	5.79 +01	5.24 +01	4.81 +01	4.44 +01	4.07 +01
16 S	1.18 +02	1.07 +02	9.64 +01	8.79 +01	8.11 +01	7.42 +01	6.71 +01	6.17 +01	5.70 +01	5.23 +01
17 CL	1.35 +02	1.23 +02	1.10 +02	1.01 +02	9.32 +01	8.52 +01	8.96 +01	7.09 +01	6.55 +01	6.01 +01
18 AR	1.56 +02	1.42 +02	1.28 +02	1.17 +02	1.08 +02	9.89 +01	8.96 +01	8.24 +01	7.62 +01	6.99 +01
19 K	2.30 +02	1.73 +02	1.55 +02	1.41 +02	1.30 +02	1.19 +02	1.07 +02	9.90 +01	9.15 +01	8.39 +01
20 CA	2.90 +02	2.08 +02	1.87 +02	1.70 +02	1.56 +02	1.43 +02	1.29 +02	1.18 +02	1.09 +02	1.00 +02
21 SC	2.51 +02	2.28 +02	2.04 +02	1.85 +02	1.72 +02	1.57 +02	1.42 +02	1.30 +02	1.20 +02	1.24 +02
22 TI	2.63 +02	2.41 +02	2.18 +02	2.01 +02	1.86 +02	1.71 +02	1.56 +02	1.44 +02	1.34 +02	1.33 +02
23 V	2.85 +02	2.61 +02	2.36 +02	2.17 +02	2.01 +02	1.85 +02	1.68 +02	1.56 +02	1.34 +02	1.61 +02
24 CR	3.52 +02	3.21 +02	2.90 +02	2.66 +02	2.46 +02	2.26 +02	2.05 +02	1.89 +02	1.75 +02	1.79 +02
25 MN	3.93 +02	3.58 +02	3.24 +02	2.96 +02	2.74 +02	2.51 +02	2.28 +02	2.10 +02	1.95 +02	1.92 +02
26 FE	3.97 +02	3.65 +02	3.33 +02	3.08 +02	2.86 +02	2.64 +02	2.41 +02	2.24 +02	2.08 +02	2.25 +02
27 CO	5.44 +02	4.94 +02	3.91 +02	3.62 +02	3.36 +02	3.10 +02	2.83 +02	2.62 +02	2.43 +02	2.32 +02
28 NI	6.20 +01	5.71 +01	5.58 +01	5.12 +01	3.36 +01	3.12 +01	2.88 +02	2.68 +02	2.50 +02	2.44 +02
29 CU	6.69 +01	6.14 +01	6.33 +01	5.81 +01	4.76 +01	4.40 +01	4.03 +01	2.82 +02	2.64 +02	2.87 +02
30 ZN	7.59 +01	6.96 +01	6.99 +01	6.44 +01	5.36 +01	4.92 +01	4.55 +01	4.22 +01	3.91 +01	4.02 +01
31 GA	8.38 +01	7.69 +01	7.25 +01	6.67 +01	5.98 +01	5.52 +01	5.05 +01	4.65 +01	4.33 +01	4.13 +01
32 GE	8.73 +01	8.00 +01	8.24 +01	7.58 +01	6.19 +01	5.71 +01	5.21 +01	4.82 +01	4.48 +01	4.74 +01
33 AS	9.89 +01	9.07 +01	8.89 +01	8.18 +01	7.04 +01	6.49 +01	5.94 +01	5.50 +01	5.13 +01	5.13 +01
34 SE	1.06 +02	9.78 +01	9.66 +01	8.89 +01	7.60 +01	7.01 +01	6.42 +01	5.95 +01	5.54 +01	5.55 +01
35 BR	1.16 +02	1.06 +02	1.02 +02	9.44 +01	8.25 +01	7.61 +01	6.96 +01	6.45 +01	6.00 +01	5.89 +01
36 KR	1.23 +02	1.13 +02	1.11 +02	1.02 +02	8.77 +01	8.08 +01	7.39 +01	6.84 +01	6.37 +01	6.37 +01
37 RB	1.34 +02	1.23 +02	1.20 +02	1.02 +02	9.52 +01	8.77 +01	8.01 +01	7.41 +01	6.90 +01	6.88 +01
38 SR	1.45 +02	1.33 +02	1.31 +02	1.10 +02	1.02 +02	9.47 +01	8.65 +01	8.00 +01	7.44 +01	7.49 +01
39 Y	1.58 +02	1.45 +02	1.41 +02	1.20 +02	1.12 +02	1.03 +02	9.42 +01	8.71 +01	8.10 +01	8.75 +01
40 ZR	1.70 +02	1.56 +02	1.50 +02	1.30 +02	1.20 +02	1.11 +02	1.01 +02	9.40 +01	8.75 +01	8.08 +01

41	NB	8.84+01	9.57+01	1.09+02	1.11+02	1.21+02	1.32+02	1.42+02	1.55+02	1.71+02	1.87+02
42	MO	9.37+01	1.01+02	1.09+02	1.18+02	1.29+02	1.40+02	1.51+02	1.65+02	1.82+02	1.99+02
43	TC	9.83+01	1.06+02	1.14+02	1.32+02	1.35+02	1.47+02	1.58+02	1.72+02	1.90+02	2.07+02
44	RU	1.05+02	1.14+02	1.23+02	1.37+02	1.49+02	1.57+02	1.70+02	1.89+02	2.08+02	2.26+02
45	RH	1.10+02	1.18+02	1.27+02	1.53+02	1.67+02	1.62+02	1.74+02	1.96+02	2.35+02	2.57+02
46	PD	1.21+02	1.31+02	1.50+02	1.69+02	1.78+02	1.93+02	2.08+02	2.13+02	2.49+02	2.72+02
47	AG	1.29+02	1.40+02	1.57+02	1.83+02	1.86+02	2.01+02	2.17+02	2.36+02	2.61+02	2.85+02
48	CD	1.35+02	1.46+02	1.74+02	1.88+02	2.06+02	2.23+02	2.40+02	2.54+02	2.79+02	3.04+02
49	IN	1.46+02	1.58+02	1.82+02	1.96+02	2.14+02	2.32+02	2.49+02	2.71+02	2.88+02	3.14+02
50	SN	1.50+02	1.62+02	1.95+02	2.10+02	2.29+02	2.48+02	2.67+02	2.90+02	2.97+02	3.24+02
51	SB	1.68+02	1.81+02	2.12+02	2.28+02	2.50+02	2.71+02	2.92+02	3.17+02	3.19+02	3.48+02
52	TE	1.82+02	1.97+02	2.38+02	2.37+02	2.59+02	2.81+02	3.03+02	3.29+02	3.50+02	3.81+02
53	I	2.05+02	2.21+02	2.49+02	2.57+02	2.81+02	3.05+02	3.28+02	3.57+02	3.63+02	3.95+02
54	XE	2.14+02	2.32+02	2.67+02	2.69+02	2.95+02	3.20+02	3.45+02	3.75+02	3.93+02	4.29+02
55	CS	2.46+02	2.52+02	2.87+02	2.96+02	3.40+02	3.50+02	3.78+02	4.33+02	4.13+02	4.51+02
56	BA	2.64+02	2.67+02	3.07+02	3.10+02	3.64+02	3.69+02	3.98+02	4.64+02	4.55+02	4.97+02
57	LA	2.82+02	2.86+02	3.42+02	3.32+02	3.92+02	3.95+02	4.26+02	5.01+02	4.77+02	5.21+02
58	CE	2.94+02	3.06+02	3.60+02	3.57+02	4.06+02	4.41+02	4.76+02	5.16+02	5.12+02	5.59+02
59	PR	3.09+02	3.18+02	3.83+02	3.70+02	4.28+02	4.65+02	5.01+02	4.83+02	5.26+02	5.02+02
60	ND	3.28+02	3.56+02	3.97+02	3.90+02	4.55+02	5.05+02	5.33+02	5.35+02	5.64+02	6.11+02
61	PM	3.40+02	3.69+02	4.23+02	4.29+02	4.67+02	4.50+02	4.60+02	4.83+02	4.94+02	5.40+02
62	SM	3.62+02	3.92+02	3.85+02	3.98+02	4.13+02	4.94+02	3.27+02	3.58+02	3.69+02	3.88+02
63	EU	3.46+02	3.75+02	2.72+02	2.79+02	4.32+02	3.16+02	3.60+02	3.33+02	3.91+02	4.02+02
64	GD	3.78+02	4.10+02	2.82+02	2.96+02	3.07+02	3.33+02	3.44+02	1.40+02	1.45+02	1.38+02
65	TB	3.94+02	4.25+02	2.07+02	2.94+02	3.18+02	3.49+02	3.28+02	1.47+02	1.53+02	1.58+02
66	DY	3.46+02	3.75+02	1.11+02	1.11+02	1.21+02	1.26+02	1.36+02	1.59+02	1.67+02	1.66+02
67	HO	2.42+02	2.68+02	1.07+02	1.19+02	1.26+02	1.31+02	1.41+02	1.64+02	1.74+02	1.81+02
68	ER	2.49+02	2.03+02	1.11+02	1.24+02	1.30+02	1.36+02	1.51+02	1.71+02	1.80+02	1.89+02
69	TM	2.60+02	3.10+02	1.15+02	1.33+02	1.36+02	1.47+02	1.57+02	1.82+02	1.87+02	1.96+02
70	YB	1.07+02	1.15+02	1.23+02	1.33+02	1.45+02	1.56+02	1.68+02	1.92+02	1.99+02	2.04+02
71	LU	1.13+02	1.22+02	1.30+02	1.40+02	1.53+02	1.65+02	1.77+02	1.95+02	2.10+02	2.18+02
72	HF	1.16+02	1.25+02	1.33+02	1.43+02	1.56+02	1.68+02	1.80+02	1.99+02	2.13+02	2.28+02
73	TA	1.19+02	1.28+02	1.37+02	1.49+02	1.61+02	1.73+02	1.85+02	2.19+02	2.17+02	2.32+02
74	W	1.22+02	1.31+02	1.39+02	1.53+02	1.60+02	1.77+02	1.86+02	2.25+02	2.42+02	2.34+02
75	RE	1.30+02	1.42+02	1.49+02	1.61+02	1.75+02	1.91+02	2.04+02	2.43+02	2.46+02	2.62+02
76	OS	1.33+02	1.40+02	1.53+02	1.65+02	1.79+02	1.94+02	2.08+02	2.52+02	2.75+02	2.67+02
77	IR	1.44+02	1.55+02	1.66+02	1.78+02	1.94+02	2.09+02	2.33+02	2.62+02	2.85+02	2.89+02
78	PT	1.50+02	1.61+02	1.72+02	1.85+02	2.01+02	2.17+02	2.43+02	2.43+02	2.85+02	2.99+02
79	AU	1.57+02	1.71+02	1.82+02	1.95+02	2.11+02	2.27+02	2.43+02	2.62+02	3.48+02	3.08+02
80	HG	1.59+02	2.08+02	2.21+02	2.38+02	2.57+02	2.77+02	2.96+02	3.19+02	3.47+02	3.75+02
81	TL	1.94+02	2.08+02	2.22+02	2.38+02	2.57+02	2.77+02	2.96+02	3.19+02	3.47+02	3.75+02
82	PB	1.94+02	2.93+02	3.17+02	3.45+02	3.81+02	4.16+02	4.51+02	4.94+02	5.50+02	6.05+02

167

MASS ATTENUATION COEFFICIENTS FOR L ALPHA 1 LINES

EMITTER	HG	TL	PB	BI	PO	AT	RN	FR	RA	AC
WAVELENGTH	1.24	1.20	1.17	1.14	1.11	1.08	1.05	1.03	1.00	.979

ABSORBER

Z	El	HG	TL	PB	BI	PO	AT	RN	FR	RA	AC
1	H	4.31 -01	4.28 -01	4.25 -01	4.22 -01	4.20 -01	4.17 -01	4.14 -01	4.11 -01	4.10 -01	4.08 -01
2	HE	2.87 -01	2.82 -01	2.76 -01	2.70 -01	2.64 -01	2.59 -01	2.53 -01	2.47 -01	2.44 -01	2.41 -01
3	LI	4.15 -01	4.00 -01	3.85 -01	3.70 -01	3.55 -01	3.40 -01	3.24 -01	3.09 -01	3.02 -01	2.95 -01
4	BE	7.52 -01	7.17 -01	6.83 -01	6.47 -01	6.12 -01	5.76 -01	5.39 -01	5.04 -01	4.87 -01	4.71 -01
5	B	1.37 +00	1.30 +00	1.23 +00	1.16 +00	1.08 +00	1.01 +00	9.35 -01	8.62 -01	8.28 -01	7.93 -01
6	C	2.23 +00	2.11 +00	1.98 +00	1.86 +00	1.73 +00	1.60 +00	1.46 +00	1.34 +00	1.28 +00	1.22 +00
7	N	2.91 +00	2.68 +00	2.46 +00	2.23 +00	2.00 +00	1.76 +00	1.53 +00	2.30 +00	2.19 +00	2.08 +00
8	O	5.77 +00	5.43 +00	5.08 +00	4.74 +00	4.38 +00	4.03 +00	3.66 +00	3.32 +00	3.16 +00	2.99 +00
9	F	7.61 +00	7.16 +00	6.71 +00	6.26 +00	5.80 +00	5.34 +00	4.87 +00	4.42 +00	4.20 +00	3.99 +00
10	NE	1.15 +01	1.08 +01	1.02 +01	9.51 +00	8.81 +00	8.09 +00	7.37 +00	6.69 +00	6.35 +00	6.02 +00
11	NA	1.43 +01	1.34 +01	1.24 +01	1.17 +01	1.09 +01	1.00 +01	9.11 +00	8.26 +00	7.85 +00	7.44 +00
12	MG	2.09 +01	1.96 +01	1.83 +01	1.71 +01	1.58 +01	1.45 +01	1.32 +01	1.19 +01	1.13 +01	1.07 +01
13	AL	2.63 +01	2.47 +01	2.31 +01	2.16 +01	1.99 +01	1.83 +01	1.67 +01	1.51 +01	1.43 +01	1.35 +01
14	SI	3.03 +01	2.85 +01	2.67 +01	2.48 +01	2.30 +01	2.11 +01	1.92 +01	1.74 +01	1.65 +01	1.56 +01
15	P	3.70 +01	3.48 +01	3.26 +01	3.03 +01	2.81 +01	2.58 +01	2.35 +01	2.12 +01	2.02 +01	1.91 +01
16	S	3.75 +01	3.48 +01	3.26 +01	3.03 +01	2.61 +01	2.80 +01	3.02 +01	2.73 +01	2.59 +01	2.45 +01
17	CL	5.75 +01	5.43 +01	4.18 +01	4.47 +01	4.04 +01	3.80 +01	3.46 +01	3.13 +01	2.98 +01	2.82 +01
18	AR	6.35 +01	5.97 +01	5.59 +01	5.21 +01	4.83 +01	4.44 +01	4.04 +01	3.66 +01	3.47 +01	3.29 +01
19	K	7.61 +01	7.15 +01	6.70 +01	6.24 +01	5.79 +01	5.31 +01	4.83 +01	4.37 +01	4.15 +01	3.93 +01
20	CA	9.10 +01	8.55 +01	8.01 +01	7.45 +01	6.89 +01	6.32 +01	5.75 +01	5.20 +01	4.93 +01	4.67 +01
21	SC	1.11 +02	1.06 +02	9.83 +01	8.22 +01	7.61 +01	6.98 +01	6.35 +01	5.75 +01	5.45 +01	5.16 +01
22	TI	1.17 +02	1.09 +02	1.08 +02	9.39 +01	7.37 +01	8.06 +01	7.33 +01	6.72 +01	6.33 +01	6.06 +01
23	V	1.22 +02	1.14 +02	1.08 +02	1.00 +02	9.73 +01	8.65 +01	7.91 +01	6.83 +01	6.86 +01	6.50 +01
24	CR	1.47 +02	1.38 +02	1.30 +02	1.21 +02	1.12 +02	1.04 +02	9.49 +01	8.57 +01	8.20 +01	7.77 +01
25	MN	1.63 +02	1.54 +02	1.44 +02	1.35 +02	1.25 +02	1.15 +02	1.05 +02	9.57 +01	9.10 +01	8.62 +01
26	FE	1.76 +02	1.66 +02	1.57 +02	1.47 +02	1.37 +02	1.26 +02	1.16 +02	1.06 +02	1.16 +02	1.17 +02
27	CO	2.06 +02	1.94 +02	1.82 +02	1.70 +02	1.58 +02	1.46 +02	1.34 +02	1.30 +02	1.24 +02	1.24 +02
28	NI	2.13 +02	2.01 +02	1.90 +02	1.78 +02	1.66 +02	1.54 +02	1.42 +02	1.37 +02	1.31 +02	1.44 +02
29	CU	2.66 +02	2.51 +02	2.36 +02	2.21 +02	2.06 +02	1.90 +02	1.75 +02	1.59 +02	1.51 +02	1.49 +02
30	ZN	2.86 +02	2.51 +02	2.53 +02	2.37 +02	2.20 +02	2.04 +02	1.87 +02	1.71 +02	1.62 +02	1.85 +02
31	GA	3.71 +02	3.45 +02	3.33 +02	3.03 +02	2.82 +02	2.60 +02	2.37 +02	2.15 +02	1.96 +02	2.51 +02
32	GE	3.81 +02	3.57 +02	3.76 +02	3.04 +02	3.54 +02	3.70 +02	3.09 +02	2.87 +02	2.69 +02	2.51 +02
33	AS	4.71 +01	4.42 +01	4.13 +01	3.64 +01	3.21 +01	3.09 +01	3.35 +01	3.10 +01	2.92 +01	2.75 +01
34	SE	5.41 +01	4.80 +01	4.81 +01	4.51 +01	4.24 +01	3.66 +01	3.59 +01	3.28 +01	3.08 +01	2.87 +01
35	BR	5.84 +01	5.52 +01	5.20 +01	4.87 +01	4.54 +01	4.20 +01	3.86 +01	3.53 +01	3.33 +01	3.13 +01
36	KR	6.30 +01	5.88 +01	5.57 +01	5.25 +01	4.93 +01	4.53 +01	4.19 +01	3.81 +01	3.61 +01	3.41 +01
37	RB	6.86 +01	5.95 +01	6.10 +01	5.71 +01	5.32 +01	4.93 +01	4.52 +01	4.14 +01	3.35 +01	3.75 +01
38	SR	7.41 +01	7.00 +01	6.59 +01	6.18 +01	5.76 +01	5.33 +01	4.90 +01	4.49 +01	4.28 +01	4.07 +01
39	Y										
40	ZR										

41	NB	8.10+01	7.65+01	7.20+01	6.75+01	6.28+01	5.82+01	5.34+01	4.89+01	4.65+01	4.43+01
42	MO	8.58+01	8.11+01	7.62+01	7.11+01	6.65+01	6.15+01	5.65+01	5.16+01	4.92+01	4.58+01
43	TC	9.01+01	8.51+01	8.01+01	7.51+01	6.99+01	6.47+01	5.95+01	5.47+01	5.19+01	5.32+01
44	RU	9.69+01	9.15+01	8.62+01	8.08+01	7.53+01	6.98+01	6.43+01	5.87+01	5.61+01	5.64+01
45	RH	1.19+02	1.05+02	8.90+01	8.29+01	7.91+01	7.34+01	6.76+01	6.21+01	5.93+01	6.09+01
46	PD	1.19+02	1.12+02	1.06+02	9.92+01	8.64+01	7.91+01	7.34+01	6.72+01	6.40+01	6.54+01
47	AG	1.34+02	1.27+02	1.19+02	1.12+02	9.25+01	8.57+01	7.98+01	7.22+01	7.14+01	6.79+01
48	CD	1.37+02	1.30+02	1.29+02	1.21+02	1.04+02	9.70+01	8.93+01	7.49+01	7.96+01	7.43+01
49	IN	1.54+02	1.46+02	1.37+02	1.39+02	1.13+02	1.11+02	9.70+01	8.19+01	8.50+01	7.57+01
50	SN	1.73+02	1.58+02	1.49+02	1.45+02	1.30+02	1.20+02	1.03+02	8.90+01	9.02+01	8.23+01
51	SB	1.88+02	1.77+02	1.67+02	1.57+02	1.35+02	1.25+02	1.15+02	9.45+01	9.70+01	9.53+01
52	TE	1.96+02	1.85+02	1.75+02	1.65+02	1.53+02	1.35+02	1.24+02	1.05+02	1.08+02	1.08+02
53	I	2.13+02	2.01+02	1.89+02	1.77+02	1.64+02	1.52+02	1.39+02	1.19+02	1.13+02	1.15+02
54	XE	2.46+02	2.28+02	2.15+02	1.88+02	1.75+02	1.62+02	1.49+02	1.27+02	1.21+02	1.23+02
55	CS	2.58+02	2.43+02	2.29+02	2.14+02	1.87+02	1.73+02	1.59+02	1.45+02	1.38+02	1.32+02
56	BA	2.69+02	2.53+02	2.38+02	2.23+02	1.99+02	1.84+02	1.64+02	1.54+02	1.46+02	1.39+02
57	LA	2.82+02	2.66+02	2.51+02	2.35+02	2.18+02	1.92+02	1.85+02	1.60+02	1.53+02	1.46+02
58	CE	3.00+02	2.83+02	2.66+02	2.49+02	2.32+02	2.02+02	1.96+02	1.79+02	1.76+02	1.67+02
59	PR	3.31+02	3.12+02	2.93+02	2.74+02	2.55+02	2.22+02	2.16+02	1.97+02	1.87+02	1.78+02
60	ND	3.46+02	3.26+02	3.06+02	2.86+02	2.66+02	2.35+02	2.27+02	2.05+02	2.06+02	1.95+02
61	PM	3.82+02	3.61+02	3.22+02	3.17+02	2.80+02	2.46+02	2.37+02	2.16+02	2.16+02	2.05+02
62	SM	3.30+02	3.58+02	3.37+02	3.16+02	2.95+02	2.72+02	2.49+02	2.27+02	2.33+02	2.14+02
63	EU	3.30+02	3.12+02	2.94+02	2.72+02	2.65+02	2.59+02	2.45+02	2.45+02	2.29+02	2.32+02
64	GD	2.20+02	2.12+02	2.08+02	1.95+02	2.05+02	2.79+02	2.65+02	2.40+02	2.51+02	2.18+02
65	TB	2.50+02	2.37+02	2.16+02	2.01+02	1.73+02	2.85+02	2.57+02	2.33+02	1.49+02	2.39+02
66	DY	2.89+02	2.77+02	2.16+02	2.06+02	1.97+02	2.33+02	2.38+02	2.21+02	1.57+02	2.08+02
67	HO	1.07+02	1.03+02	9.73+01	8.74+01	9.99+01	1.83+02	1.69+02	1.60+02	1.68+02	1.56+02
68	ER	1.13+02	1.07+02	1.09+02	9.08+01	8.41+01	1.86+02	1.80+02	1.69+02	1.26+02	1.61+02
69	TM	1.22+02	1.15+02	1.10+02	1.02+02	8.93+01	1.93+02	8.28+01	7.72+01	8.25+01	6.80+01
70	YB	1.28+02	1.16+02	1.10+02	1.04+02	9.62+01	8.95+01	8.38+01	3.37+01	8.04+01	7.42+01
71	LU	1.33+02	1.26+02	1.24+02	1.17+02	1.05+02	8.30+01	9.08+01	3.46+01	9.11+01	7.79+01
72	HF	1.38+02	1.31+02	1.32+02	1.25+02	1.17+02	8.06+01	3.02+01	1.14+01	1.11+02	8.04+01
73	TA	1.70+02	1.40+02	1.61+02	1.52+02	1.42+02	9.82+01	1.23+02	1.42+02	1.25+02	8.61+01
74	W	1.80+02	1.71+02	1.32+02	1.53+02	1.44+02	1.33+02	1.25+02	1.14+02	1.16+02	1.06+02
75	RE	2.44+02	2.30+02	2.15+02	2.01+02	1.86+02	1.71+02	1.54+02	1.42+02	1.35+02	1.27+02

MASS ATTENUATION COEFFICIENTS FOR L ALPHA 1 LINES
EMITTER   TH   PA   U

WAVELENGTH	.955	.932	.910
ABSORBER			
1 H	4.07 -01	4.05 -01	4.04 -01
2 HE	2.39 -01	2.36 -01	2.33 -01
3 LI	2.88 -01	2.81 -01	2.74 -01
4 BE	4.54 -01	4.36 -01	4.19 -01
5 B	7.58 -01	7.22 -01	6.86 -01
6 C	1.16 +00	1.10 +00	1.03 +00
7 N	1.97 +00	1.86 +00	1.75 +00
8 O	2.83 +00	2.66 +00	2.49 +00
9 F	3.77 +00	3.54 +00	3.32 +00
10 NE	5.68 +00	5.33 +00	4.99 +00
11 NA	7.02 +00	6.59 +00	6.16 +00
12 MG	1.01 +01	9.48 +00	8.85 +00
13 AL	1.28 +01	1.20 +01	1.12 +01
14 SI	1.47 +01	1.38 +01	1.29 +01
15 P	1.80 +01	1.69 +01	1.58 +01
16 S	2.31 +01	2.17 +01	2.03 +01
17 CL	2.65 +01	2.49 +01	2.32 +01
18 AR	3.10 +01	2.91 +01	2.72 +01
19 K	3.70 +01	3.47 +01	3.24 +01
20 CA	4.40 +01	4.12 +01	3.85 +01
21 SC	4.87 +01	4.57 +01	4.26 +01
22 TI	5.73 +01	5.38 +01	5.04 +01
23 V	6.14 +01	5.77 +01	5.40 +01
24 CR	7.33 +01	5.89 +01	6.44 +01
25 MN	8.13 +01	7.64 +01	7.14 +01
26 FE	9.08 +01	8.55 +01	8.02 +01
27 CO	1.04 +02	9.82 +01	9.19 +01
28 NI	1.11 +02	1.05 +02	9.86 +01
29 CU	1.18 +02	1.11 +02	1.04 +02
30 ZN	1.36 +02	1.28 +02	1.20 +02
31 GA	1.45 +02	1.36 +02	1.27 +02
32 GE	1.41 +02	1.33 +02	1.25 +02
33 AS	1.75 +02	1.64 +02	1.53 +02
34 SE	1.76 +02	1.66 +02	1.56 +02
35 BR	2.57 +01	2.39 +01	1.66 +02
36 KR	2.71 +01	2.56 +01	2.40 +01
37 RB	2.92 +01	2.71 +01	2.55 +01
38 SR	3.21 +01	3.01 +01	2.80 +01
39 Y	3.55 +01	3.35 +01	3.14 +01
40 ZR	3.85 +01	3.63 +01	3.41 +01

41	NB	4.19+01	3.96+01	3.72+01		
42	MO	4.43+01	4.17+01	3.92+01		
43	TC	4.67+01	4.40+01	4.14+01		
44	RU	5.03+01	4.75+01	4.46+01		
45	RH	5.35+01	5.06+01	4.76+01		
46	PD	5.76+01	5.43+01	5.10+01		
47	AG	6.20+01	5.85+01	5.50+01		
48	CD	6.43+01	6.07+01	5.70+01		
49	IN	7.17+01	6.77+01	6.36+01		
50	SN	7.67+01	7.25+01	6.82+01		
51	SB	8.13+01	7.68+01	7.22+01		
52	TE	8.74+01	8.25+01	7.75+01		
53	I	9.09+01	8.57+01	8.06+01		
54	XE	9.80+01	9.25+01	8.69+01		
55	CS	1.02+02	9.64+01	9.06+01		
56	BA	1.09+02	1.03+02	9.66+01		
57	LA	1.17+02	1.10+02	1.03+02		
58	CE	1.24+02	1.17+02	1.10+02		
59	PR	1.31+02	1.24+02	1.16+02		
60	ND	1.38+02	1.30+02	1.22+02		
61	PM	1.45+02	1.36+02	1.28+02		
62	SM	1.53+02	1.44+02	1.35+02		
63	EU	1.58+02	1.49+02	1.40+02		
64	GD	1.68+02	1.58+02	1.48+02		
65	TB	1.76+02	1.65+02	1.55+02		
66	DY	1.85+02	1.74+02	1.63+02		
67	HO	1.94+02	1.83+02	1.71+02		
68	ER	1.96+02	1.84+02	1.73+02		
69	TM	2.03+02	1.91+02	1.79+02		
70	YB	2.09+02	1.97+02	1.85+02		
71	LU	2.20+02	2.07+02	1.94+02		
72	HF	2.06+02	1.95+02	1.83+02		
73	TA	2.26+02	2.13+02	2.00+02		
74	W	2.32+02	2.19+02	2.07+02		
75	RE	2.32+02	2.20+02	2.08+02		
76	OS	2.03+02	1.90+02	2.14+02		
77	IR	1.49+02	1.89+02	1.73+02		
78	PT	1.51+02	1.42+02	1.32+02		
79	AU	1.50+02	1.41+02	1.36+02		
80	HG	1.53+02	1.45+02	1.50+02		
81	TL	1.59+02	1.59+02	1.53+02		
82	PB	6.94+01	6.83+01	7.30+01		
83	BI	7.31+01	7.74+01	8.94+01		
86	RN	8.18+01	7.45+01	9.07+01		
90	TH	9.97+01	9.59+01	1.06+02		
92	U	1.01+02	9.13+01			
94	PU	1.20+02	1.13+02			

APPENDIX

2

AVERAGE VALUES OF FLUORESCENT YIELDS

The values given in Tables A2.1 and A2.2 were determined from the curves plotted by Fink et al.*

*R. W. Fink, R. C. Jopson, H. Mark, and C. D. Swift, *Rev. Mod. Phys.*, **38**, 513 (1966).

Table A2.1. Average K Fluorescent Yields

Z	Element	ω_K	Z	Element	ω_K	Z	Element	ω_K
6	C	0.0009	24	Cr	0.26	41	Nb	0.755
7	N	0.0015	25	Mn	0.285	42	Mo	0.77
8	O	0.0022	26	Fe	0.32	43	Tc	0.785
10	Ne	0.0100	27	Co	0.345	44	Ru	0.80
11	Na	0.020	28	Ni	0.375	45	Rh	0.81
12	Mg	0.030	29	Cu	0.41	46	Pd	0.82
13	Al	0.040	30	Zn	0.435	47	Ag	0.83
14	Si	0.055	31	Ga	0.47	48	Cd	0.84
15	P	0.070	32	Ge	0.50	49	In	0.85
16	S	0.090	33	As	0.53	50	Sn	0.86
17	Cl	0.105	34	Se	0.565	51	Sb	0.87
18	Ar	0.125	35	Br	0.60	52	Te	0.875
19	K	0.140	36	Kr	0.635	53	I	0.88
20	Sc	0.165	37	Rb	0.665	54	Xe	0.89
21	Ca	0.190	38	Sr	0.685	55	Cs	0.895
22	Ti	0.220	39	Y	0.71	56	Ba	0.90
23	V	0.240	40	Zr	0.72			

Table A2.2. Average L Fluorescent Yields[a]

Z	Element	ω_L	Z	Element	ω_L	Z	Element	ω_L
40	Zr	0.057	59	Pr	0.168	77	Ir	0.340
41	Nb	0.061	60	Nd	0.173	78	Pt	0.353
42	Mo	0.067	61	Pm	0.178	79	Au	0.363
43	Tc	0.073	52	Sm	0.183	80	Hg	0.373
44	Ru	0.080	63	Eu	0.190	81	Tl	0.382
45	Rh	0.085	64	Gd	0.196	82	Pb	0.391
46	Pd	0.091	65	Tb	0.203	83	Bi	0.399
47	Ag	0.096	66	Dy	0.208	84	Po	0.405
48	Cd	0.103	67	Ho	0.213	85	At	0.410
49	In	0.109	68	Er	0.220	86	Ru	0.417
50	Sn	0.115	69	Tm	0.227	87	Fr	0.423
51	Sb	0.120	70	Yb	0.237	88	Ra	0.428
52	Te	0.125	71	Lu	0.252	89	Ac	0.431
53	I	0.131	72	Hf	0.268	90	Th	0.436
54	Xe	0.138	73	Ta	0.285	91	Pa	0.440
55	Cs	0.145	74	W	0.302	92	U	0.443
56	Ba	0.152	75	Re	0.315	93	Np	0.445
57	La	0.157	76	Os	0.327	94	Pu	0.448
58	Ce	0.163						

[a] The values are given to 2 or 3 places to distinguish adjacent elements but any individual value is only accurate to perhaps 5 to 10%.

APPENDIX

3

JUMP FACTORS FOR SELECTED ELEMENTS

The tabulated K-edge jump values are from the curve of Fig. A.1 which shows the jump factors listed in the Compilation of X-ray Cross Sections report UCRL-50174 Sect. II, Rev. 1 by W. H. McMaster, N. Kerr del Grande, J. H. Mallett, J. H. Hubbell, May 1969 Lawerence Radiation Laboratory, Univ. of Cal., Livermore, Cal. The L-edge jump factors are obtained in a similar fashion. Errors of 5 to 10% may be present but since the jump factor only enters into the secondary fluorescence correction term of eq. 7.13, the resulting uncertainty in composition is usually less than 1%.

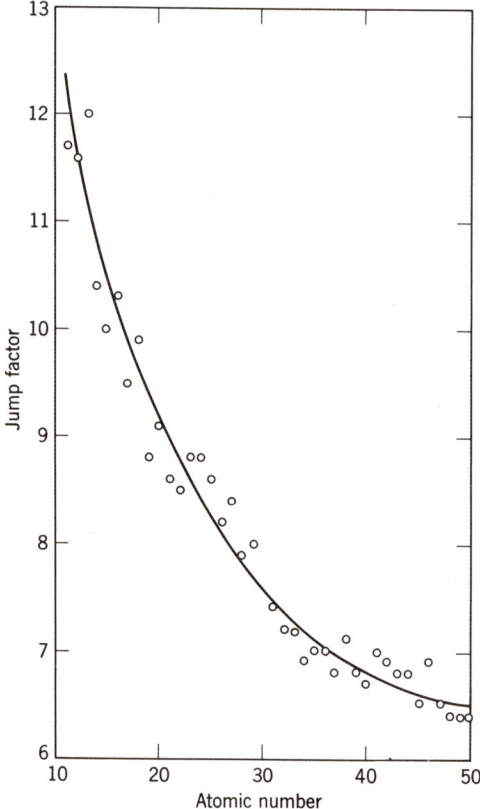

Fig. A3.1 K-edge jump factor.

Table A3.1. *K*-Edge Jump Factors

Element	K jump factor	Element	K jump factor
11 Na	12.3	32 Ge	7.4
12 Mg	11.8	33 As	7.3
13 Al	11.1	34 Se	7.2
14 Si	10.8	35 Br	7.1
15 P	10.5	36 Kr	7.0
16 S	10.2	37 Rb	7.0
17 Cl	9.8	38 Sr	6.9
18 Ar	9.6	39 Yt	6.9
19 K	9.3	40 Zr	6.8
20 Ca	9.1	41 Nb	6.8
21 Sc	8.9	42 Mo	6.7
22 Ti	8.8	43 Tc	6.7
23 V	8.6	44 Ru	6.6
24 Cr	8.4	45 Rh	6.6
25 Mn	8.3	46 Pd	6.6
26 Fe	8.1	47 Ag	6.6
27 Co	8.0	48 Cd	6.5
28 Ni	7.9	49 In	6.5
29 Cu	7.7	50 Sn	6.5
30 Zn	7.6		
31 Ga	7.5		

Table A3.2. *L*-Edge Jump Factor

Element	L jump factor	Element	L jump factor
50 Sn	5.0	78 Pt	3.7
51 Sb	4.9	79 Au	3.7
52 Te	4.8	80 Hg	3.7
53 I	4.7	81 Tl	3.7
54 Xe	4.6	82 Pb	3.6
55 Cs	4.5	83 Bi	3.6
56 Ba	4.5	84 Po	3.6
57 La	4.4	85 At	3.6
60 Nd	4.3	90 Th	3.5
65 Tb	4.1	91 Pa	3.5
70 Yb	4.0	92 U	3.5
72 Hf	3.9		
73 Ta	3.9		
74 W	3.9		
75 Re	3.8		
76 Os	3.8		
77 Ir	3.8		

APPENDIX

4

EXCITATION ENERGY FOR *K*, *L*, AND *M* SERIES[a]

Element	Z	K	L_1	M_1
C	6	.28 k.e.v.		
N	7	.40		
O	8	.53		
F	9	.69		
Na	11	1.1		
Mg	12	1.3	.06 k.e.v.	
Al	13	1.6	.09	
Si	14	1.8	.12	
P	15	2.1	.15	
S	16	2.5	.19	
Cl	17	2.8	.24	
A	18	3.2	.29	
K	19	3.6	.34	
Ca	20	4.0	.40	
Sc	21	4.5	.46	
Ti	22	5.0	.53	
V	23	5.5	.60	
Cr	24	6.0	.68	
Mn	25	6.5	.76	
Fe	26	7.1	.85	.09 k.e.v.
Co	27	7.7	.93	.10
Ni	28	8.3	1.0	.11
Cu	29	9.0	1.1	.12
Zn	30	9.7	1.2	.14
Ga	31	10.4	1.3	.18
Ge	32	11.1	1.4	.19
As	33	11.8	1.5	.20
Se	34	12.7	1.7	.23
Br	35	13.4	1.8	.26
Kr	36	14.3	1.9	.29
Rb	37	15.2	2.1	.32
Sr	38	16.1	2.2	.36
Y	39	17.1	2.4	.41
Zr	40	18.0	2.5	.43
Nb	41	19.0	2.7	.47
Mo	42	20.0	2.9	.51

Element	Z	K	L_1	M_1
Tc	43	21.1	3.0	.55
Ru	44	22.1	3.2	.58
Rh	45	23.2	3.4	.62
Pd	46	24.4	3.6	.68
Ag	47	25.5	3.8	.74
Cd	48	26.7	4.0	.78
In	49	27.9	4.2	.82
Sn	50	29.2	4.4	.87
Sb	51	30.5	4.7	.95
Te	52	31.8	5.0	1.0
I	53	33.2	5.2	1.1
Xe	54	34.6	5.5	1.1
Cs	55	36.0	5.7	1.2
Ba	56	37.4	6.0	1.3
La	57	——	6.3	1.4
Ce	58		6.6	1.4
Pr	59		6.8	1.5
Nd	60		7.1	1.6
Pm	61		7.4	1.7
Sm	62		7.7	1.7
Eu	63		8.1	1.8
Gd	64		8.4	1.9
Tb	65		8.7	2.0
Dy	66		9.1	2.0
Ho	67		9.4	2.1
Er	68		9.8	2.2
Tm	69		10.1	2.3
Yb	70		10.5	2.4
Lu	71		10.9	2.5
Hf	72		11.3	2.6
Ta	73		11.7	2.7
W	74		12.1	2.8
Re	75		12.5	2.9
Os	76		13.0	3.0
Ir	77		13.4	3.2
Pt	78		13.9	3.3
Au	79		14.4	3.4
Hg	80		14.8	3.6
Tl	81		15.3	3.7
Pb	82		15.9	3.9
Bi	83		16.4	4.0
Po	84		16.9	4.2
At	85		17.5	4.3
Fr	87		18.7	4.6

(*continued*)

Element	Z	K	L_1	M_1
Ra	88		19.2	4.8
Ac	89		19.8	5.0
Th	90		20.5	5.2
Pa	91		21.1	5.4
U	92		21.8	5.5
Np	93		22.4	5.7
Pu	94		23.1	5.9
Am	95		23.8	6.1
Cu	96		24.5	6.3
Be	97		25.2	6.6
Cf	98		26.0	6.8
E	99		26.8	7.0
Fm	100		27.1	7.3

[a]Data for this table were taken from *The Handbook of Chemistry and Physics*, 43rd ed., The Chemical Rubber Publishing Co., Cleveland, Ohio, 1961, or extrapolated by the author.

[b]K-series excitation is not recommended for electron probe analysis above 38 k.e.v.

APPENDIX

5

CHARACTERISTIC WAVELENGTHS OF THE K-, L-, AND M- SERIES LINES[a]

Element	Z	K series		L series		M series	
		$\alpha_{1,2}$	$\beta_{1,2}$	α_1	β_1	α_1	β_1
C	6						
N	7						
O	8						
F	9						
Na	11	11.9 A.	11.6				
Mg	12	9.9	9.5				
Al	13	8.3	8.0				
Si	14	7.1	6.8				
P	15	6.1	5.8				
S	16	5.4	5.0				
Cl	17	4.7	4.4				
K	19	3.8	3.4				
Ca	20	3.4	3.1				
Sc	21	3.0	2.8				
Ti	22	2.74	2.51				
V	23	2.50	2.28	24.3 A.			
Cr	24	2.29	2.08	21.5	21.2		
Mn	25	2.10	1.90	19.4	19.0		
Fe	26	1.93	1.75	17.6	17.2		
Co	27	1.79	1.62	15.9	15.6		
Ni	28	1.66	1.50	14.5	14.2		
Cu	29	1.54	1.39	13.3	13.0		
Zn	30	1.43	1.29	12.2	12.0		
Ga	31	1.34	1.20	11.3	11.0		
Ge	32	1.25	1.12	10.4	10.1		
As	33	1.17	1.05	9.7	9.4		
Se	34	1.10	.99	9.0	8.7		
Br	35	1.04	.93	8.4	8.1		
Rb	37	.92	.82	7.5	7.3		
Sr	38	.87	.78	6.8	6.6		
Y	39	.83	.73	6.4	6.2		
Zr	40	.78	.70	6.1	5.8		

(*continued*)

Element	Z	K series		L series		M series	
		$\alpha_{1,2}$	$\beta_{1,2}$	α_1	β_1	α_1	β_1
Nb	41	.75	.66	5.7	5.5		
Mo	42	.71	.63	5.4	5.2		
Ru	44	.64	.57	4.8	4.6		
Rh	45	.61	.54	4.6	4.4		
Pd	46	.59	.52	4.4	4.1		
Ag	47	.56	.49	4.1	3.9		
Cd	48	.53	.47	3.9	3.7		
In	49	.51	.45	3.8	3.5		
Sn	50	.49	.43	3.6	3.4		
Sb	51	.47	.41	3.4	3.2		
Te	52			3.28	3.07		
I	53			3.14	2.93		
Cs	54			2.89	2.68		
Ba	56			2.77	2.56		
La	57			2.66	2.45		
Ce	58			2.56	2.35		
Pr	59			2.46	2.25		
Nd	60			2.37	2.16		
Sm	62			2.20	1.99	11.4 A.	
Eu	63			2.12	1.92	10.9	
Gd	64			2.04	1.83	10.4	
Tb	65			1.97	1.77	10.0	9.8
Dy	66			1.90	1.71	9.5	9.3
Ho	67			1.84	1.64	9.1	8.9
Er	68			1.78	1.58	8.8	8.6
Tm	69			1.72	1.53		
Yb	70			1.67	1.47	8.1	7.9
Lu	71			1.62	1.42	7.8	7.6
Hf	72			1.57	1.37	7.5	7.3
Ta	73			1.52	1.32	7.2	7.0
W	74			1.47	1.28	7.0	6.7
Re	75			1.43	1.24	6.7	6.5
Os	76			1.39	1.19	6.5	6.3
Ir	77			1.35	1.16	6.2	6.0
Pt	78			1.31	1.18	6.0	5.8
Au	79			1.27	1.08	5.8	5.6
Hg	80			1.24	1.05	5.6	5.4
Tl	81			1.20	1.01	5.5	5.2
Pb	82			1.17	0.98	5.3	5.1
Bi	83			1.14	.95	5.1	4.9
Th	90			.95	.76	4.1	3.9
U	92			.91	.72	3.9	3.7
Np	93			.89	.70	3.7	3.5

Element	Z	K series		L series		M series	
		$\alpha_{1,2}$	$\beta_{1,2}$	α_1	β_1	α_1	β_1
Pu	94			.87	.68	3.5	3.4
Am	95			.85	.66	3.4	2.4
Cu	96			.83	.64	3.3	3.1
Be	97			.81	.62	3.2	3.0
Cf	98			.79	.60	3.0	2.9
E	99			.78	.59	2.9	2.8
Fm	100			.76	.57	2.8	2.6

[a] Data for the tables were taken from:
1. *Handbook of Chemistry and Physics*, 43rd ed., The Chemical Rubber Publishing Co., Cleveland, Ohio, 1961.
2. M. Siegbahn, *Spektroskopie der Rontgenstrahlen*, Springer, Berlin, 1931.
3. A. H. Compton and S. K. Allison, *X-Rays in Theory and Experiment*, 2nd ed., van Nostrand, New York, 1943.
4. Extrapolation by the author.

[b] K-series radiation is not recommended for electron probe analysis above atomic number 51.

APPENDIX

6

USEFUL ANALYZER CRYSTALS AND THEIR SPACINGS

Table A6.1

Crystal	Planes	d spacing
LiF	(200)	2.01
EDDT[a]	(020)	4.38
Aluminum	(111)	2.31
NaCl	(200)	2.81
KCl	(200)	3.14
KBr	(200)	3.29
Quartz	($10\bar{1}1$)	3.35
ADP[b]	(200)	3.75
Quartz	($10\bar{1}0$)	4.25
Pentaerythritol	(002)	4.38
ADP[b]	(110)	5.31
Oxalic acid	(001)	5.85
Graphite	(001)	6.69
Mica	(002)	9.9
Silver acetate	(001)	10.0

[a] Ethylenediamine d-tartrate.
[b] Ammonium dihydrogen phosphate.

The Bragg angle θ is obtained from the equation $n\lambda = 2d \sin \theta$ where λ is the wavelength in A.

For an excellent tabulation of wavelength, the reader is referred to *Handbuch der Physik,* Volume 30. Shorter tabulations that are sufficient for much practical work may be found in *The Handbook of Chemistry and Physics,* any volume.

REFERENCES

1. G. von Hevesy, *Chemical Analysis by X-Rays and its Application*, McGraw-Hill, New York, 1932; reprinted by University Microfilms Inc., Ann Arbor, Mich., 1960.
2. (a) M. von Ardenne, *Elektronenubermikroskopie*, Springer, Berlin, 1940. (b) V. K. Zworykin, G. A. Morton, E. G. Ramberg, J. Hillier, and A. W. Vance, *Electron Optics and the Electron Microscope*, Wiley, New York, 1945.
3. (a) M. von Ardenne, *Z. Physik*, **109**, 533 (1938). (b) K. C. A. Smith and C. W. Oatley, *Brit. J. Appl. Phys.*, **6**, 391 (1955).
4. J. Hillier, U.S. Pat. 2,418,029 (1947).
5. R. Castaing and A. Guinier, *Electron Microscope*, Proc. Delft Conf., 1949, p. 60.
6. R. Castaing, Ph.D. Thesis, University of Paris, 1951.
7. (a) I. B. Borovskii, *Problemy Met. (Moscow)*, **1953**, 135; (b) I. B. Borovskii and N. P. Il'in; *Dokl. Akad. Nauk SSSR*, **106**, 655 (1953). (c) I. B. Borovskii and N. P. Il'in, *Exp. Tech.*, **5**, 36 (1957).
8. V. E. Cosslett and P. Duncumb, *Nature*, **177**, 1172 (1956).
9. M. E. Haine and T. Mulvey, *J. Sci. Instr.*, **36**, 350 (1959).
10. L. S. Birks and E. J. Brooks, *Rev. Sci. Instr*, **28**, 709 (1957).
11. R. M. Fisher and J. C. Schwartz *Proc. Conf. Ind. Appl. X-Ray Anal.*, 5th Denver Conf., 1956; *J. Appl. Phys.*, **28**, 1377 (1957).
12. D. B. Wittry, Dept. of the Army Contract DA-04-495-Ord-463, *Repts. No. WAL 142/59-5, 142/59-6;* Thesis, California Institute of Technology, 1957.
13. C. S. Schwartz and A. E. Austin, *J. Appl. Phys.*, **28**, 1368 (1957).
14. D. B. Langmuir, *Proc. IRE*, **25**, 977 (1937).
15. R. Castaing, *Laboratories*, **17**, 7 (1956).
16. R. Castaing, *Advan. Electron. Electron Phys.*, **13**, 317 (1961).
17. T. Mulvey, *Brit. J. Appl. Phys.*, **8**, 259 (1957) (abstract only); *J. Sci. Instr.*, **36**, 350 (1959); *Electron Microscope*, Proc. Berlin Conf., 1958.
18. D. B. Wittry, "Instrumentation for Electron Probe Analysis," *ARL Report*, September 1959.
19. J. B. Le Poole, *Proc. 3rd. European Conf. on Electron Microscope*, Prague, Czech. Acad. Sci., 1964, p. 439.
20. P. Duncumb, *Proc. 5th International Congress on Electronmicroscopy* (paper KK4), Academic Press, New York and London, 1962.
21. G. Liebmann, *Proc. Phys. Soc. (London)*, **B68**, 737 (1955).
22. S. K. Allison, *Phys. Rev.*, **41**, 1 (1932).
23. L. S. Birks and R. T. Seal, *J. Appl. Phys.*, **28**, 541 (1957).
24. J. Vierling, J. V. Gilfrich, and L. S. Birks, *Appl. Spectroscopy*, **23**, 342 (1969).

25. L. Hailes, *Conference of the Electron Microscopy and Analysis Group,* Institute of Physics, London, February 1969.
26. R. W. Gould, S. R. Bates, C. J. Sparks, *Appl. Spectroscopy,* **22,** 549 (1968).
27. T. Johansson, *Naturwissenschaften,* **20,** 758 (1932).
28. H. H. Johann, *Z. Physik,* **69,** 185 (1931).
29. Y. Cauchois, *J. Phys.,* **3,** 329 (1932).
30. S. A. Ditsman, *Bull. Acad. Sci. USSR, Phys. Ser. (Eng. Transl.),* **24,** 390 (1960).
31. L. S. Birks, R. E. Seebold, A. P. Batt, and J. S. Grosso, *J. Appl. Phys.,* **35,** 2578 (1964).
32. A. H. Compton and S. K. Allison, *X-rays in Theory and Experiment,* van Nostrand, New York, 1935, p. 740.
33. L. S. Birks, R. E. Seebold, B. K. Grant, and J. S. Grosso, *J. Appl. Phys.,* **36,** 699 (1965).
34. D. W. Aitkin, *IEEE Trans. Nucl. Sci.,* **NS15** (3), 10, (1968).
35. K. F. J. Heinrich, D. Vieth, H. Yakowitz; *Adv. X-Ray Anal.* **9,** 208 (1966).
36. G. Dearnaley and D. C. Northrop, *Semiconductor Counters for Nuclear Reactions,* Wiley, New York, 1963, Chap. 6.
37. R. E. Ogilvie and R. Lewis, "Measurement of Diffusion in an Al-Zn Diffusion Couple," *8th Ann. Conf. Appl. X-ray Anal.,* Denver, Colorado, 1959.
38. R. M. Dolby, *Proc. Phys. Soc.,* **73,** 81 (1959).
39. L. S. Birks, R. J. Labrie, and J. W. Criss, *Anal. Chem.,* **38,** 701 (1965).
40. R. Castaing, *Advan. Electron. Electron Phys.,* **13,** 317 (1960).
41. I. Adler, *ASTM Spec. Tech. Pub. 349,* p. 183, 1963.
42. J. Rucklidge and E. F. Stumpfl, *Neues Jahrbuch f. Mineralogie,* Monatshefte, 1-2 (1968), 61, I. Adler, *Symposium on X-Ray and Electron Probe Analysis,* ASTM Special Technical Publication No. 349 (1964), 184, M. P. Borom and R. E. Hannerman, *2nd Nat'l. Conf. on Electron Microprobe Anal.,* Boston, Massachusetts, 1967.
43. A. J. Tousimis, *ASTM Spec. Tech. Pub. 349,* p. 193, 1963.
44. C. A. Anderson, K. Keil, and B. Mason, *Science,* **146,** 256 (1964).
45. L. S. Birks, J. M. Siomkajlo, and P. K. Koh, *Trans. AIME,* **218,** 806 (1960).
46. L. S. Birks and A. P. Batt, *Anal. Chem.,* **35,** 778 (1963).
47. R. M. Fisher, *J. Appl. Phys.,* **24,** 113 (1953).
48. P. Duncumb, *The Electron Microprobe,* T. D. McKinley, K. F. J. Heinrich, and D. B. Wittry, Eds., Wiley, New York, 1966, p. 490.
49. N. L. Peterson and R. E. Ogilvie, "Electron Microbeam Probe," *AAAS Annual Meeting,* Chicago, December 1959.
50. T. Mamuro, A. Fujita, T. Matsunami, S. Shirai, and M. Murakami, *Ann. Rept. of the Radiation Center of Osaka Prefecture,* **6,** 14 (1965).
51. M. R. Achter, L. S. Birks, and E. J. Brooks, *J. Appl. Phys.,* **30,** 1825 (1959).
52. D. J. Nagel, U.S. Naval Research Laboratory, unpublished work.
53. R. E. Seebold and L. S. Birks, *Anal. Chem.,* **34,** 112 (1962.)

54. L. S. Birks, unpublished work given at 12th International Colloquium Spectroscopy, Exeter, Great Britain, September 1965.
55. W. E. Sweeney, R. E. Seebold, and L. S. Birks, *J. Appl. Phys.*, **31**, 1061 (1960).
56. J. Kimoto and J. C. Russ, *Am. Sci.*, **57**, 112 (1969).
57. L. Mayer, *J. Appl. Phys.*, **29**, 658 (1958).
58. R. M. Dolby, *J. Sci. Instr.*, **40**, 345 (1963).
59. K. Keil and C. A. Anderson, *Geochim. Cosmochim. Acta*, **29**, 621 (1965).
60. J. Fica, E. I. DuPont, Wilmington, Delaware, private communication.
61. L. S. Birks, E. J. Brooks, I. Adler, and C. Milton, *Am. Mineralogist*, **44**, 974 (1959).
62. J. W. Criss and L. S. Birks, *Anal. Chem.*, **40**, 1080 (1968).
63. R. Castaing, Ph.D. Thesis, University of Paris, 1951.
64. Dr. Beaman and J. A. Isasi, *Anal. Chem.*, **42**, 1540 (1970); Readers in search of current computer programs are advised to contact the secretary of the Electron Probe Analysis of America.
65. J. Henoc, "Quantitative Electron Probe Microanalysis," K. F. J. Heinrich, Ed., *Nat. Bur. Std., Spec. Publ. 298*, 1968.
66. J. Henoc, F. Maurice, A. Kirianenko, Centre d'Études Nucléaires de Saclay, France, *Report CEA-R 2421* (1964).
67. P. Duncumb, S. J. B. Reed, "Quantitative Electron Probe Microanalysis," K. F. J. Heinrich, Ed., *Nat. Bur. Std. Spec. Publ. 298*, 1968.
68. J. Philibert, *X-Ray Optics and X-Ray Microanalysis*, H. H. Pattee, V. E. Cosslett, A. Engström, Eds., Academic Press, New York, 1963, p. 379.
69. P. Duncumb and P. K. Shields, *Brit. J. Appl. Phys.*, **14**, 617 (1963).
70. S. J. B. Reed, *Brit. J. Appl. Phys.*, **16**, 913 (1965).
71. K. F. J. Heinrich, *2nd Natl. Conf. on Electron Microprobe Anal.*, Boston, Massachusetts, 1967.
72. P. Duncumb, P. K. Shields-Mason, and C. da Casa, *5th International Congress on X-Ray Optics and Microanalysis*, Tübingen, Germany, 1968, G. Mollenstedt and K. H. Gaukler, Eds., Springer-Verlag, Berlin, Heidelberg, New York, 1969, p. 146.
73. J. Criss, "Quantitative Electron Probe Microanalysis," K. F. J. Heinrich, Ed., *Natl. Bur. Std. Spec. Publ. 298*, 1968, p. 53.
74. J. W. Criss and L. S. Birks, *The Electron Microprobe*, T. D. McKinley, K. F. J. Heinrich, and D. B. Wittry, Eds., Wiley, New York, 1966, p. 217.
75. J. H. Wegstein, *Commun. Assoc. Computer Machinery*, **1**, 9 (1958).
76. S. B. J. Reed, *2nd Natl. Conf. on Electron Microprobe Anal.*, Boston, Massachusetts, June 1967.
77. D. B. Brown and R. E. Ogilvie, *J. Appl. Phys.*, **37**, 4429 (1966).
78. D. B. Brown, D. B. Wittry, and D. F. Kyser, *J. Appl. Phys.*, **40**, 1627 (1969).
79. See for example H. A. Bethe, *Handbuch für Physik*, Springer, Berlin, 1933 Vol. 24/1, p. 273.
80. F. Rohrlich and B. C. Carlson, *Phys. Rev.*, **93**, 38 (1954).

81. M. J. Berger and S. M. Seltzer, *Natl. Acad. Sci. Publ. 1133,* Washington, D.C. 1964, p. 205.
82. H. A. Bethe, M. E. Rose, and L. P. Smith, *Proc. Amer. Phil. Soc.,* **78,** 573 (1938).
83. R. W. Fink, R. C. Jopson, H. Mark, and C. D. Swift, *Rev. Mod. Phys.,* **38,** 513 (1966).
84. D. B. Brown, *5th International Conf. on X-Ray Optics and Microanal.,* Tübingen, Germany, September 1968 (to be published separately from the conference proceedings).
85. M. Green, *Proc. Phys. Soc. (London),* **82,** 204 (1963).
86. H. E. Bishop, *X-ray Optics and Microanalysis,* R. Castaing, et al., Eds., Hermann, Paris, 1966, p. 112.
87. K. Murata, R. Shimizu, and G. Shinoda, *Tech. Rep. Osaka Univ.,* **16,** 121 (1966).
88. L. S. Birks, D. J. Ellis, B. K. Grant, A. S. Frisch, and R. B. Hickman, *The Electron Microprobe,* T. D. McKinley K. F. J. Heinrich, D. B. Wittry, Eds., Wiley, New York, 1966, p. 199.
89. M. J. Berger, *Natl. Bur. Std.,* Washington, D.C., unpublished work.
90. R. Castaing and G. Slodzian, *9th International Colloquium Spectroscopy,* Lyons, France, 1961.
91. F. W. Karasek, *Res. Devel.,* **21,** 32 (1970).
92. H. Liebl, *J. Appl. Phys.,* **38,** 5277 (1967).
93. D. M. Hercules, *Anal. Chem.,* **42,** 20A (1970) (review paper).
94. D. Betteridge and A. D. Baker, *Anal. Chem.,* **42,** 43A (1970) (review paper).
95. E. J. Brooks, Naval Research Laboratory, unpublished work.
96. S. Kimoto and J. C. Russ, *Amer. Sci.,* **57,** 112 (1969).
97. G. P. Thompson and W. Cochran, *Electron Diffraction,* MacMillan, London, 1939, p. 112.
98. E. Bauer, *Techniques of Metals Research,* R. F. Bunshah, Ed., Wiley-Interscience, 1969, p. 559 (review paper).
99. S. Kikuchi, *Proc. Japan Acad.,* **4,** 534 (1928).
100. W. Kossel, *Nachr. Akad, Wiss Goettingen, II, Math-Physik Kl.,* **1,** 229 (1935).
101. M. v. Laue, *Ann. Physik,* **25,** 569 (1936).
102. Z. G. Pinsker, *Electron Diffraction,* Eng. transl. by J. A. Spink and E. Feigl, Butterworth, London, 1953, p. 40ff.
103. K. Lonsdale, *Phil. Trans. Roy. Soc., London,* **240,** 219 (1946–48).
104. V. E. Cosslett and W. C. Nixon, *X-ray Microscopy,* Cambridge University Press, 1960.
105. P. Duncumb, *The Electron Microprobe,* T. D. McKinley et al., Eds., Wiley, New York, 1966, p. 490.
106. J. B. LePoole, *Proc. 3rd. European Conf. on Electron Microscopy,* Prague, Czech. Acad. Sci. 1964, p. 439.

AUTHOR INDEX

Achter, M. R., 90
Adler, I., 72, 103
Aitkin, D. W., 58
Allison, S. K., 42, 53
Anderson, C. A., 82, 99
Ardenne, M. V., 6, 7
Austin, A. E., 13

Baker, A. D., 133
Banner, E., 137
Batt, A. P., 50, 83
Beaman, D., 106
Berger, M. J., 125, 129
Bethe, H., 125, 126
Betteridge, D., 133
Birks, L. S., 12, 44, 50, 53, 68, 83, 90, 91, 94, 103, 104, 119, 129
Bishop, H. E., 129
Borom, M. P., 72
Borovskii, I. B., 9
Bracewell, B., 147
Briggs, E., 147
Brooks, E. J., 12, 90, 103, 135
Brown, D. B., 109, 124, 128

Carlson, B. C., 125
da Casa, C., 114
Castaing, R., 8, 26, 71, 105, 130
Cauchois, Y., 46
Cockran, W., 136, 140
Compton, A. H., 53
Cosslett, V. E., 10, 144
Criss, J. W., 68, 104, 116, 119

Dearnley, C., 64
Ditsman, S. A., 49
Dolby, R. M., 68, 99
Donaldson, M., 147
Duncumb, P., 10, 29, 86, 113, 114, 145

Ellis, D. J., 129

Ficca, J., 100
Fink, R. W., 128, 172
Fisher, R. M., 12, 84
Frisch, A. S., 129

Fujita, A., 87

Gilfrich, J. V., 44
Gould, R. W., 44
Grant, B. K., 53, 129
Green, M., 129
Grosso, J., 50, 53
Guinier, A., 8

Hailes, L., 44
Haine, M. E., 10
Hannerman, R. E., 72
Heinrich, K. F. J., 63, 113
Henoc, J., 112
Hercules, D. M., 133
Hevesy, G. v., 5
Hickman, R. B., 129
Hitlier, J., 6–8

Il'in, N. P., 9
Isasi, J. A., 106

Johann, H. H., 45
Johansson, T., 45
Jopson, R. C., 128, 172

Karasek, F. W., 130
Keil, K., 82, 99
Kikuchi, S., 140
Kimoto, J., 97, 135
Kirianenko, A., 112
Koh, P. K., 83
Kossel, W., 140
Kyser, D. F., 109, 124, 128

Labrie, R. J., 68
Langnuir, D. B., 23
Laue, M. v., 143
LePoole, J. P., 29, 145
Lewis, R., 66
Liebl, H., 133
Liebman, G., 33
Lonsdale, K., 143

Mamuro, T., 87
Mark, H., 128, 172

Mason, B., 82
Matsunami, T., 87
Maurice, F., 112
Mayer, L., 97
McMaster, W., 174
Millon, C., 103
Morton, G. A., 6, 7
Mulvey, T., 10, 27
Murakami, M., 87
Murata, K., 129

Nagel, D. J., 99
Nixon, W. C., 144
Northrop, D. C., 64

Oatley, C. U., 7
Ogilvie, R., 66, 86, 124

Peterson, N. L., 86
Philibert, J., 113
Pinsker, Z. G., 143

Ramberg, E. G., 6, 7
Reed, S. J. B., 113, 120, 128
Rohrlich, F., 125
Rose, M. E., 126
Rucklidge, J., 72
Russ, J. C., 97, 135

Schwartz, C. S., 13

Schwartz, J. C., 12
Seal, R. T., 44
Seebold, R. E., 50, 53, 93, 94
Seltzer, S. M., 125
Shields, P. K., 113, 114
Shimizu, R., 129
Shinoda, G., 129
Shirai, S., 87
Siomkajlo, J. M., 83
Slodzian, G., 130
Smith, K. C. A., 7
Smith, L. P., 126
Strumpfl, E. F., 72
Sweeney, W. E., 94
Swift, C. C., 128, 172

Thompson, G. P., 136, 140
Tousimis, A. J., 75, 79

Vance, A. W., 6, 7
Veigele, W. J., 147
Vierling, J., 44
Vieth, D., 63

Wegstein, J. H., 120
Wittry, D. B., 12, 27, 109, 124, 128

Yakowitz, H., 63

Zworykin, V. K., 6, 7

SUBJECT INDEX

Aberration, astigmatism, 35–36
  coma, 35–36
  lens, 31–37
  spherical, 32–35
Absorption factor, 113
Alignment, column, 37–40
Analysis, quantitative, 101ff, 124ff
Aperture, 24, 35
Astigmatism, 35–36
Atomic number correction, 113–115
Auger electron, 134

Backscatter, electron, 106–108, 114
Beam current, 35
Beam deflection, 30–31
Beam size, 34
Bias resistor, 22
  voltage, 22
Bragg angle, 41–42
Brightness, 35

Calibration curves, 101
Cathodoluminescence, 99
Chi, 111
Coma, 35–36
Commercial manufacturers, electron probe, 15
  electron probe-microscope, 146
  ion probe, 130
Conductivity, electrical, 72
Contamination spot, 20
Correction factors, 105
  absorption, 113
  atomic number, 113–115
  fluorescence, 116
  numerical example, 116–119
Counting rate, 59
Crystal optics, 45–47
Crystal parameters, 42–45, 182
Curved-crystal optics, 45–47

Dead time, 57, 63–64
Detectability, 4, 70
Detector, electron, 64–66
  visible, 66
  x-ray, 57–63

Diffraction, electron, 136
  x-ray, 41–42, 138
Diffusion, 87–91
Dispersion, energy, 66–70
  wavelength, 41ff
Display, 30, 80–82, 97, 99–100, 131–132
Divergence, fanning, 47

Electron diffraction, 136
Electron excitation, 109–111
  gun, 21
  lens, 23–30
  microscopy, 6, 29–30, 135–136, 145–146
  optics, 20ff
  range, 107
  scattering, 126
  spectrometer, 133–135
  stopping power, 114
  transport, 124–128
    limitations, 128–129
Empirical coefficients, 104
Energy, x-ray, 176
  x-ray excitation, 176ff
Energy dispersion, 66–70
Energy spectra, 68
Enhancement, 112, 116, 122
Errors, 120–123
Excitation, electron, 109–111
Excitation energy, x-ray, 176
Extraction replica, 83–85

Ferromagnetic specimens, 26
Fluorescence, characteristic, 116
  secondary, 112, 123
Fluorescence correction, 116
Fluorescence yield, 116, 172–173
Focal length, 32
Focusing circle, 45

History of electron probe development, 5–8

Instrument design, 7–15
Intensity, background, 3
  beam, 35
  line/background, 53–54
  x-ray, 50–51, 109–111

Ion probe, 130–133
Ionization potential, 114–115
Iteration, 119–120

Jump factor, 116, 174–175

K-edge jump ratio, 116, 174
Kikucni pattern, 140–141
Kossel pattern, 142–143

L-edge jump ratio, 175
Line/background ratio, 53–54

Magnetic field strength, 26
Mass absorption coefficient, 147ff
Matrix effect, *see* Correction factors
Microscopy, scanning, 135–136
  transmission, 6, 29–30, 144–146
  x-ray, 144–145
Monte Carlo calculations, 123–124, 129
Multichannel analyzer, 67

Oil, vacuum pump, 20
Overvoltage, 116

Particulate speciments, 86
Photoelectron, 134
Pole piece design, 25–29
Precipitates, 80
Proportional counter, 57–61
Pulse amplitude, 67
Pulse height, 58

Range, electron, 107
Regression equations, 104
Resolution, electron beam, 23, 33
  energy, 57–59, 66
  wavelength, 6, 48–49, 52–53

Scanning, 55–56; *see also* Display
Scanning microscopy, 135–136
Scintillation counter, 61–62
Secondary fluorescence, 112, 122
Size, electron beam, 32–35
  x-ray source, 108
Specimen, biological, 75
  cutting, 75
  diffusion, 87
  hard, 71
  inclusion, 80
  mounting, 76
  particulate, 86
  polishing, 71–72
  preparation, 71ff
  soft, 75
  thin film, 92
  types, 72
Spectra, x-ray, 179ff
Spectrometer, electron, 133
  x-ray, 41ff
Spectrometer mechanics, 54–55
Spherical aberration, 32–35
Stabilization, current, 23
  voltage, 23
Standards, 78
  comparison, 102
Stopping power, electron, 114
Substrate, 75
Surface roughness, 71

Take-off angle, 71
Temperature, specimen, 71
Thin films, 92

Unfolding, spectral, 67–69

Vacuum requirements, 20

Wavelengrh, dispersion, 41ff
  x-ray, 5, 179ff
X-ray absorption coefficients, 147ff
  analysis, 101ff, 124ff
  crystals, 42–45, 182
  detector, 57–63
  diffraction, 41–42, 138
  display, 30, 80–82, 97, 99–100, 131–132
  energy, 68
  fluorescence, 116
  intensity, 50–51, 109–111
  jump factor, 116, 174–175
X-ray microscopy, 144–145
  source size, 108
  spectra, 179ff
  spectrometer, 41ff
  wavelength, 5, 179ff

/543.085B619E1971>C1/